Planet Earth
ARID LANDS

Other Publications:

FIX IT YOURSELF
FITNESS, HEALTH & NUTRITION
SUCCESSFUL PARENTING
HEALTHY HOME COOKING
UNDERSTANDING COMPUTERS
LIBRARY OF NATIONS
THE ENCHANTED WORLD
THE KODAK LIBRARY OF CREATIVE PHOTOGRAPHY
GREAT MEALS IN MINUTES
THE CIVIL WAR
COLLECTOR'S LIBRARY OF THE CIVIL WAR
THE EPIC OF FLIGHT
THE GOOD COOK
WORLD WAR II
HOME REPAIR AND IMPROVEMENT
THE OLD WEST

For information on and a full description of any of
the Time-Life Books series listed above, please write:
   Reader Information
   Time-Life Books
   541 North Fairbanks Court
   Chicago, Illinois 60611

This volume is one of a series that examines the
workings of the planet earth, from the geological
wonders of its continents to the marvels of its
atmosphere and its ocean depths.

Cover
Against a backdrop of petrified sandstone in
northeastern Arizona, part of the Great
American Desert, the bladelike leaves of a yucca
plant emerge from a dune of sand. The yucca
survives the drying glare of the sun by storing
water in special cells beneath its tough skin.

Planet Earth

ARID LANDS

By Jake Page
and The Editors of Time-Life Books

Time-Life Books, Alexandria, Virginia

Time-Life Books Inc.
is a wholly owned subsidiary of

TIME INCORPORATED

FOUNDER: Henry R. Luce 1898-1967

Editor-in-Chief: Henry Anatole Grunwald
Chairman and Chief Executive Officer: J. Richard Munro
President and Chief Operating Officer: N. J. Nicholas Jr.
Chairman of the Executive Committee: Ralph P. Davidson
Corporate Editor: Ray Cave
Executive Vice President, Books: Kelso F. Sutton
Vice President, Books: George Artandi

TIME-LIFE BOOKS INC.

EDITOR: George Constable
Executive Editor: Ellen Phillips
Director of Design: Louis Klein
Director of Editorial Resources: Phyllis K. Wise
Editorial Board: Russell B. Adams Jr., Thomas H.
Flaherty, Lee Hassig, Donia Ann Steele, Rosalind
Stubenberg, Kit van Tulleken, Henry Woodhead
Director of Photography and Research:
John Conrad Weiser

PRESIDENT: Christopher T. Linen
Chief Operating Officer: John M. Fahey Jr.
Senior Vice Presidents: James L. Mercer,
Leopoldo Toralballa
Vice Presidents: Stephen L. Bair, Ralph J. Cuomo,
Neal Goff, Stephen L. Goldstein, Juanita T. James,
Hallett Johnson III, Robert H. Smith,
Paul R. Stewart
Director of Production Services: Robert J. Passantino

PLANET EARTH

SERIES DIRECTOR: Thomas A. Lewis
Deputy Editor: Russell B. Adams Jr.
Designer: Albert Sherman
Chief Researcher: Patti H. Cass

Editorial Staff for *Arid Lands*
Associate Editor: Marion F. Briggs (pictures)
Text Editors: Sarah Brash, Jan Leslie Cook,
Thomas H. Flaherty
Researchers: Blaine Marshall, Barbara Moir
(principals), Barbara Brownell, Roxie M. France,
Sara Mark
Assistant Designer: Cynthia T. Richardson
Copy Coordinator: Elizabeth Graham
Picture Coordinator: Renée DeSandies
Editorial Assistant: Mary Kosak

Special Contributors: Champ Clark, Karen Jensen,
Donna Roginski (text)

Editorial Operations
Copy Chief: Diane Ullius
Editorial Operations Manager: Caroline A. Boubin
Production: Celia Beattie
Quality Control: James J. Cox (director)
Library: Louise D. Forstall

Correspondents: Elisabeth Kraemer-Singh (Bonn);
Maria Vincenza Aloisi (Paris); Ann Natanson
(Rome). Valuable assistance was also provided by:
Angelika Lemmer (Bonn); Robert Kroon (Geneva);
Marlin Levin, Jean Max (Jereusalem); Lesley
Coleman, Millicent Trowbridge (London); Carolyn
T. Chubet, Christina Lieberman (New York);
Ann Wise (Rome).

Library of Congress Cataloguing in Publication Data
Page, Jake.
  Arid lands.
  (Planet earth)
  Bibliography: p.
  Includes index.
  1. Arid regions. 2. Desert ecology. I. Time-Life
Books. II. Title. III. Series.
GB611.P26 1984   910'.02'154   83-18105
ISBN 0-8094-4512-3
ISBN 0-8094-4513-1 (lib. bdg.)

THE AUTHOR

Jake Page is a contributing editor and columnist
for *Science '84*. He was formerly the editor of
Natural History magazine and has served as science editor of *Smithsonian* magazine. He has
written extensively about environmental sciences
and natural history, and he is the author of *Forest*, another volume in the Planet Earth series.

THE CONSULTANTS

Dr. Farouk El-Baz established the Center for
Earth and Planetary Studies at the Smithsonian
Institution's National Air and Space Museum
and served as the center's director from 1973 to
1982. In addition, between 1978 and 1981, he
served as Science Adviser to the late President
Anwar Sadat of Egypt. A fellow of the Geological Society of America, the American Association for the Advancement of Science and
the Royal Astronomical Society, Dr. El-Baz has
published more than 200 scientific papers and
six books, including *Apollo-Soyuz Test Project:
Earth Observations and Photography* and *Egypt As
Seen by Landsat*.

Frederic H. Wagner is Associate Dean of the
College of Natural Resources, Director of the
Ecology Center and Professor of Fisheries and
Wildlife at Utah State University. He has studied, written and lectured about the ecology and
resource management of arid lands. He has spent
time in the deserts of every continent and is best
known for his book, *Wildlife of the Deserts*.

Peter L. Kresan was named Instructor of Physical and Historical Geology at the University of
Arizona in 1981. He has been an interpretative
naturalist, guide and photographer for museum
and private tours throughout the Southwest.
His interests include landforms and geomorphic
processes in desert environments.

CONTENTS

THE AUSTERE BEAUTY OF THE DESERT

Unlike any other environment on earth, a desert is defined by what it lacks: water, soil, vegetation and population. By that measure, the beautiful blue planet is a far less benign place than might be imagined. Indeed, fully one third of the earth's land surface is classified as arid, much of it dun-colored wastelands; what is more, these arid lands are increasing by an estimated 63 square miles every day in response to the present workings of the earth.

In general terms, a desert is an area that receives less than 10 inches of precipitation a year. Territories that qualify include not only blazing reaches of the North African Sahara where no rain has fallen for 20 years but the icebound continent of Antarctica and coastal regions of Chile that are almost perpetually shrouded in fog.

The desert floor is littered with the unassembled constituents of soil: sand, dust, pebbles and rock. (The sandy dunes that symbolize the desert exist in only a fraction — 12 per cent — of the lands classified as arid.) Without moisture to bind the particles together and nurture plant life, the barren expanses are endlessly battered by wind and tortured by extremes of hot and cold that are found nowhere else on the planet. This relentlessly hostile environment is home to only the hardiest of living things — plants whose seeds can endure a 50-year drought, for example, and rodents that live their entire lives without a drink of water.

Citizens of nearly 70 nations confront the spreading deserts, here falling back in famine, there pressing forward to deliver to the stricken land the blessings of moisture denied by nature. As with all enterprises that seek to alter the effects of titanic global forces, the contest is extremely difficult and expensive. And although accommodations are possible, the desert, for all it lacks, will endure.

Swirling dunes ripple across the Namib Desert on the southwestern coast of Africa. Tireless southerly winds erode the area's reddish bedrock into sand, then pile the sand in ribbons that reach heights of 1,200 feet.

Deeply weathered into rockfalls and sand by the harsh desert environment, ancient cliffs tower over one of five Bande-e Amir lakes 10,000 feet above sea

level in the Hindu Kush mountains of Afghanistan. Variations in the composition of the underlying bedrock color each lake a rich hue.

Rounded boulders three feet in diameter dot a limestone plateau in Chad's Tibesti Massif. In a little-understood phenomenon, extreme temperature

fluctuations common to high-altitude deserts, along with fierce winds, shape the stones by eroding surrounding layers of sedimentary rock.

A lake too salty to freeze mirrors walls of weathered rock in the dry Wright Valley of Antarctica. Although Antarctica is almost entirely covered by ice,

annual precipitation is so meager that the continent qualifies as a desert — as proved by the barren valleys where mountains block off the glaciers.

Snow softens the desolate landscape of Monument Valley, Arizona, where towering sandstone buttes preside over miles of desert. Buttes, among the

oldest of desert landforms, are the remnants of plateaus that have been eroded away by hundreds of millennia of rain and wind.

A lavender blanket of owl's-clover brightens an expanse of the Sonoran Desert in southwestern Arizona. Awakened from dormancy by a late-winter

rain, the blossoms will produce seed and die in a few weeks, before the onset of summer's desiccating heat.

DISCOVERING THE DEMONS OF NOON

Richard Trench had eagerly looked forward to his venture into the enormous expanse of the Sahara. The young English journalist had carefully plotted the route that would take him more than 950 cruel, parched miles from Chegga, a desert village in northeastern Mauritania, south to Timbuktu, the legendary "Queen of the Desert" on the Niger River. Trench had purchased his equipment, laid in his supplies and made arrangements to travel with a group of Bedouin cameleers. Most of all, he had immersed himself in the lore and literature of the early desert explorers, all dead now, the bones of some lying bleached or buried in the endless sand.

But nothing that Richard Trench had done or read could prepare him for the insensate violence he experienced on this afternoon in the fall of 1974.

He was returning on foot to his encampment near Chegga from a little oasis about two miles away, where the Bedouins' camels were grazing. A mild breeze cooled his face; the great glaring orb of the sun, usually so stunning in its assault upon the human senses, now smoldered sullen behind a yellowish haze. And then, without other warning, the storm struck.

"Suddenly the wind began to rise, blowing in short and powerful gusts," Trench recalled. "The surface of the desert, normally so still, was growing restless. As the wind rose, so the desert rose too. It was dancing about my feet. It hurled itself against my calves, whirled around my body and beat at my bare arms. Nor did it stop there. It grabbed my neck, stung my face, encrusted itself in my throat, blocked my nostrils and blinded my eyes. I felt alone and feared death by drowning."

Blasted from his feet by the shrieking wind, Trench lay curled in the sand — which soon threatened to bury him. Remembering that there was a nearby escarpment, he crawled toward the shelter it might provide. Groping in the darkness of the storm, he saw an indefinable object to his left. Struggling toward it on his hands and knees, he was almost atop it before he realized that it was the corpse of a camel: "Sand was piled high against its hump, its skin hung like parchment from its semi-exposed ribs, and its insides were swollen into vile and bloated shapes."

Trench crawled on, still seeking the escarpment. But by the time he reached it, "the tempest had calmed down. The air was still heavy with dust, but the ground was beginning to look firm again. The wind had dropped and the heavy particles of sand were floating down onto the desert floor in layers. Everything was very peaceful."

Richard Trench, who would successfully complete his journey to Timbuktu, had just seen the desert in one of its many moods. As time is reck-

Remarkably distinct in the clear sky, the moon illuminates a timeless scene of camels and riders atop the dunes of the Thar Desert in northwest India. Since the 12th Century, caravans have crossed this desert bearing loads of wool, hides, salt and other trade goods, often traveling by night, when the unbearable daytime temperatures of 120° F. or more drop by half.

oned in the ageless sands, the storm had lasted only an instant. In its fury it had churned up millions of tiny particles, but the face of the desert remained unchanged and unchangeable. And in its coming, Trench had shared an experience known by the explorers who had led the way, not only into the Sahara but the world's other great deserts—perhaps the most mysterious, sometimes alluring, often malevolent places on the surface of the earth.

Until fairly recently a desert was, by simple rule of thumb, defined as any area receiving 10 inches or less of annual rainfall. Yet convenient though it was, the criterion was often misleading: for example, many more than 10 inches may pour down in one or two sudden deluges of such volume that the earth cannot absorb the water, which runs off in flash floods, leaving the desert almost as dry as before.

To provide a more realistic measurement, modern scientists have devised a system that compares the solar energy an area receives to its annual precipitation. Thus, in the eastern Sahara and the Peruvian desert, the two driest places on earth, the sun's energy can evaporate 200 times the amount of rainfall received in an average year. The aridity index is therefore 200, and both regions are classed as hyperarid. At the far end of the scale, the Great Plains east of the American Rocky Mountains have an index that

The fierce "sand wind" for which the Sahara is notorious assaults a caravan south of Tripoli. The colored lithograph is based on an 1819 sketch by the British explorer George F. Lyon, one of the first Europeans to attempt to plumb the interior of the vast desert.

ranges from 1.5 to four; this region is termed semiarid and is capable of supporting a great diversity of life.

Of the world's biggest and most arid deserts, a great majority lies within two belts that girdle the earth near the Equator. In the Northern Hemisphere, the arid belt stretches eastward along the Tropic of Cancer from the immense Sahara of North Africa, across the Arabian Peninsula, to the forbidding Gobi and other deserts of central Asia. It includes the arid lands of the American Southwest, which are picturesque but relatively small and benign. The southern belt lies along the Tropic of Capricorn, where land masses are fewer and farther between; it includes the high plateau of southern Africa's Kalahari Desert, the extraordinarily parched deserts of Peru and Chile, and the wild Australian Outback.

Since the morning of mankind, these vast and desolate tracts have been settled or traveled by peoples who somehow eked out the bare sustenance of life not only for themselves but for their gaunt goats, their cattle or their camels. Particularly in the Arab world, there were great scholars as well, men of learning who recorded what they saw and what they knew. Such a one was Ibn Battuta, who during the span of his years in the 14th Century traveled an estimated 75,000 miles in North Africa, Arabia, Asia Minor, China and the South Seas. Battuta crossed the Sahara to Timbuktu, almost dying of thirst on the same caravan route that Richard Trench traveled six centuries later. Battuta's celebrated chronicle, *The Journey,* intrigued readers in the Islamic world and in Europe.

From the 15th Century onward, Europeans became the world's most aggressive explorers, driven in part by an insatiable urge to learn more about the world in which they lived. Europe is the only continent that is entirely lacking in arid lands, a fact that rendered accounts of distant deserts by Battuta and others all the more mysterious — and therefore appealing.

To be sure, the earliest Europeans to penetrate the deserts were driven less by love of the unknown than by the interests of business, both clerical and secular. Heathens presumably awaited conversion by Christian missionaries; and trade routes might be established to short-cut the costly, time-consuming transport of goods by sea. To those pioneer travelers, the desert was little more than an unpleasant place that had to be crossed — and the quicker the better.

By the 19th Century, however, the monks and the merchants had given way to new breeds. Some, to whom hardship was a credential of courage and danger an elixir of life, were drawn to the desert by little more than their passion for adventure. Others — and their numbers increased as the Age of Inquiry rose to its zenith — were scholars seeking their answers in the ruins of civilizations long since abandoned to desert encroachments, or scientists using their disciplines to dispel the mysteries of the vast empty spaces. And with the coming of these researchers, the deserts, slowly and painfully yet inexorably, began to give up their secrets.

Far from being utterly barren, as was once believed, the arid lands were found to contain an amazing diversity of plants and animals that have evolved ecological strategies for survival in the face of climatic and geologic conditions frightful beyond imagining. In their fascinating adaptations they not only offer testimony to the tenacity of life on earth but provide clues to the past — and perhaps to the future — of all living things.

Similarly, sophisticated Europeans once viewed the desert peoples with

an admixture of disdain and fear, conceiving them to be, at best, fierce primitives who stole by choice and would cut a strange throat for the sheer delight of the deed. Later learning has demonstrated that the inhabitants of the arid lands, rich in their own customs and cultures, in fact represent a triumph of man's dauntless ingenuity, with lessons to be taught to humans wherever they may exist in natural adversity.

With the quantum advances of the 20th Century, the world's great deserts have become laboratories for the entire range of earth sciences, and the dynamics of those natural forces that shape and govern the arid lands are integral pieces in the mosaic that may yet explain the origins of the planet — and its destiny.

It all began with the European explorers who followed the desert sun.

Far from being of a stupefying sameness composed of nothing more than undulating infinities of sun-baked sand, the deserts of earth are of bewildering variety. Some are characterized by numbing cold, with temperatures below freezing throughout much of the year. Such are the deserts of central Asia, including the Gobi, a 500,000-square-mile expanse of mountains and desolate tablelands with an average elevation of 4,000 feet. Beyond the Gobi to the west lies a series of smaller deserts, and then, stretching across the Tarim River basin in northwest China almost to the border of Pakistan, lies the formidable Taklamakan Desert. Somewhat surprisingly — more by inadvertence than design — it was in these frigid precincts that European travelers first trekked.

Across the Gobi early in the 13th Century swept the mounted Mongol hordes of Genghis Khan, on their way to conquering Asian and European territory more than twice the size of the Roman Empire at its greatest. By midcentury, rumors had reached Pope Innocent IV that Kuyuk Khan, a successor to Genghis, was ripe for conversion to Christianity; and to investigate that intriguing possibility the Pontiff sent a Franciscan friar, John of Pian de Carpine, to call upon Kuyuk at his capital in northern Mongolia.

Friar John's mission was of course foredoomed to failure; and in history's retrospect it is perhaps more significant that to get where he was going he became the first European to cross even a small part of the Gobi. Although his 30-page report to the Pope of the journey made meager mention of the Gobi, Carpine did make clear that he had not enjoyed his desert experience: "We were traveling the whole winter, resting most of the time in the snow in the desert, and often when the wind drifted in we would find, on waking, our bodies all covered with snow."

Only a quarter of a century passed before another European party ventured into the frostbitten deserts of central Asia. The brothers Nicolo and Maffeo Polo were merchants of Venice seeking trade with Kublai Khan. By happy circumstance they took with them Nicolo's 17-year-old son, Marco — whose journey, spent mostly in service as an aide to the Great Khan, would last for 24 years.

On the way to Kublai's court in what is now Beijing, the Polos struck into the Pamir, an arid, two-mile-high plateau (its inhabitants call it "the roof of the world") at the edge of the Taklamakan Desert. According to Marco Polo, they traveled for 12 days without sighting "any green thing, so that travelers are obliged to carry with them whatever they have need of. The region is so lofty and cold that you do not even see any birds flying.

A cave sculptured from a sandstone cliff by water and wind still shelters an 80-room structure erected in the 11th Century by the Anasazi Indians, one of the early cultures to cope successfully with the semiarid American Southwest. Pioneer photographer Timothy O'Sullivan took this picture of the ruin during a U.S. Army mapping expedition in 1873.

23

And I must notice also that because of the great cold, fire does not burn so brightly, nor give out so much heat as usual, nor does it cook food so effectively." (In fact, the fires lacked oxygen at that high altitude.)

At length the Venetians came to the edge of the Gobi, where even Marco Polo was somewhat daunted after being told that "it would take a year and more to ride from one end of it to the other." In the event, they traversed the narrowest end of the grim desert, an effort that nonetheless took them 30 days and 380 miles of hard going. Along the way, Marco Polo carefully counted 28 places where travelers could obtain the Gobi's most precious commodity — water. And such is the timeless nature of the desert that 600 years later the Hungarian-born British explorer Sir Aurel Stein, crossing the Gobi by the same route, also found precisely 28 water holes.

Gobi means "pebbly plain" in Mongolian, and the greater part of the desert consists of high plateaus that were, eons ago, stripped of the cover of loess — a tawny, powdery soil — by shrieking northwesterly winds that left a bare, gravelly rock surface. Yet just as the world's deserts differ greatly from one another, so does each within its own environs possess a contrasting array of land features. The Gobi is certainly no exception — as a young British Army officer named Francis E. Younghusband would learn during his travels in central Asia.

Considering the hardships that lay ahead, Younghusband set out with an astoundingly casual attitude. In 1886, after completing a temporary assignment in Beijing, Younghusband decided that it might be interesting to return to his regiment in India by crossing the Gobi. "I had never been in a desert," he wrote later, "and here were a thousand or so miles to be crossed. Nor had we any information about the state of the country on the other side." No matter. With the arrival of spring, Younghusband embarked from the Gobi's eastern margin with a guide, two porters, eight camels and a supply of sherry to quench his thirst when water was unavailable.

Foolish the trip may have been — but Francis Younghusband was no fool. Rather, he proved to be a careful, competent and, most of all, observant traveler. And in his published account of the expedition, which was remarkably free of the tall tales told by many another desert adventurer, he depicted as well as anyone the Gobi in most of its many faces.

Predictably, during most of the journey the Gobi lived up to its name, with mile after dreary mile where there existed "not a bush, nor a plant, nor a blade of grass — absolutely nothing but gravel." This presented an unexpected hazard: The sandstorms encountered on many other deserts are bad enough, but on at least one occasion Younghusband's little party was pelted by "small pebbles being driven before the wind with great velocity, which hurt us considerably."

Passing between parallel ranges of barren hills that "presented a most fantastic appearance, rising in sharp rugged peaks with intervening strips of plain, perhaps a quarter of a mile wide," Younghusband witnessed an intermediate stage of the natural process by which the Gobi's gravel was formed. "The hills of the Gobi," he noted, "are perfectly bare, and in such an extremely dry climate, exposed to the icy cold winds of winter and the fierce rays of the summer sun, and unprotected by one atom of soil, the rocks first decompose, and then crumble away to a remarkable extent."

Progressing in a westerly direction, Younghusband encountered a dramatically different topography. Rising abruptly out of the gravel plain was

The Salt Mines of "Creation's Hell Hole"

Were it not for the blazing desert heat and the passing of frequent, plodding camel caravans, the bleak Danakil Depression of northeastern Ethiopia could pass for a polar wasteland. But the white crystals that reflect the remorseless sun, creating a blinding brightness across the 2,000-square-mile plain, are composed not of ice but of salt. The vast and thirsty salt plain is the bed of Lake Assale, all that remains of what was once an arm of the Red Sea. Over the ages, volcanic eruptions cut the inlet off from the sea, and the sun repeatedly evaporated the water from the salt lake that remained, accumulating salt deposits that are in places more than three miles thick.

The same intense heat that makes the salt available makes the mineral essential to the survival of desert dwellers. The human body requires about eight ounces of salt to carry on a number of biochemical processes, such as the transmission of impulses through the nervous system. But vital organs must also be kept from overheating, and the perspiration that battles the effects of desert heat also causes a constant — and dangerous — loss of the body's salt. The need to replace it has made salt one of the desert's most valued commodities and salt mining one of its oldest trades.

The mining of the enormous salt resources of the Danakil Depression continues today as it has for millennia: as a constant struggle with the ferocity of a land that has earned the ominous sobriquet "Creation's Hell Hole." Local tribesmen use poles to pry from the lake bed huge slabs of salt that they then cut and smooth into smaller bricks. The bricks are sold to traders whose caravans arrive in a constant, year-round stream.

Lumbering across flat plains shimmering with heat and languorous mirages, the caravans transport the precious cargo 75 miles to a salt market in Makale. There, for a penny, the traders, who come from all across northeastern Africa, may nibble a chunk of salt as they haggle for the market's wares.

A caravan of salt traders from the Ethiopian highlands trudges across glistening Lake Assale to buy bricks of the vital commodity from Danakil miners.

Working in heat that frequently exceeds 125° F., Danakil salt miners in Ethiopia use poles to lever huge slabs from the dry bed of Lake Assale.

A loaded camel caravan heads through a desert mirage toward the foothills of the Ethiopian escarpment on its way to the salt market in Makale. By the

Camels wait nearby as workers cut and smooth the slabs into smaller, more manageable bricks for sale to salt traders.

time the salt reaches consumers beyond the highlands, it may bring 30 times the price paid by the caravaneers.

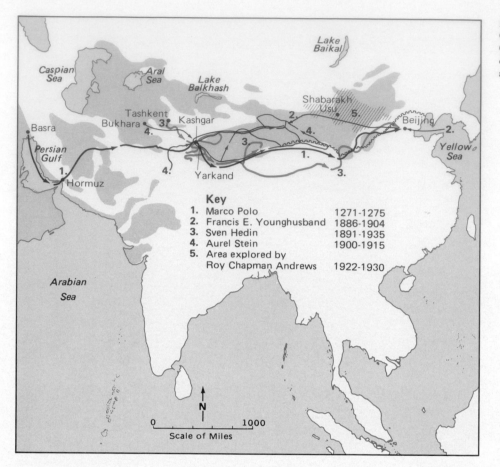

Inspired by Marco Polo's travels in the 13th Century, explorers such as Sven Hedin and Roy Chapman Andrews were lured into the Asian deserts during the following centuries. The map at left traces their epic adventures.

Key

1. Marco Polo — 1271-1275
2. Francis E. Younghusband — 1886-1904
3. Sven Hedin — 1891-1935
4. Aurel Stein — 1900-1915
5. Area explored by Roy Chapman Andrews — 1922-1930

a 40-mile-long range of hills "composed of bare sand, without a vestige of vegetation of any sort." To Younghusband, it was evident that the hills, some of them nearly 900 feet high, had been formed by sand blown there from elsewhere. The theory was corroborated when he learned that to the west lay an immense sandy tract where, according to Mongol legend, a Tartar army forming to invade China was entirely buried beneath the sands churned up by a mighty storm.

On and on marched Younghusband and his companions — "mere specks on that vast expanse of desolation" — and then, incredibly, they found themselves in danger of being stranded in the midst of a bog. Still on the move at about 11 o'clock one dark night, they came upon an area where, wrote Younghusband, "there had been a heavy rain during the day; the soil was a very slimy clay, and the ground broken up into hillocks." Suddenly the camels began floundering and, by the light of a lantern, Younghusband could see that the animals "were each perched up on a little hillock, separated from each other by pools of water and slimy clay."

From these tiny islands the beasts refused to budge. Younghusband and his companions pulled at the animals' nose strings "till I thought their whole noses would be pulled off." The camels stayed where they were. Their masters beat them, and still they balked. Finally, and with a strength born of desperation, the travelers hauled off the camels — backward. "This had the desired effect," Younghusband later recalled. "They were started, and once they were in motion they kept going."

In early July, Francis Younghusband peered into the distance and saw a pair of poplars arising from the desert plain. The next day he arrived in the

A portrait of Swedish explorer Sven Hedin dressed in winter gear hints at his fierce confidence, which was thoroughly tested by his adventures in the searing heat and bitter cold of the deserts of central Asia. Forced in 1895 to abandon two dying camels, Hedin later sketched the scene, showing one camel watching the departing caravan "with a wistful and reproachful look."

town of Hami, on the edge of the Gobi, after traveling 1,255 miles in 70 days. His journey was barely half done. After resting for four days, he set off westward again into the equally forbidding Taklamakan Desert, crossing its 900-mile expanse in a little under a month before turning south to India.

Hard upon the heels of Younghusband, the talented amateur explorer, came others of more professional — and scientific — disciplines. One of these was Sven Hedin, a Swedish geographer who, beginning in 1891, devoted more than 40 years to surveying and charting the deserts of central Asia and, by the time he completed his last expedition at the age of 71, had removed the label of Terra Incognita from the maps representing much of the territory. Another was the scholarly Sir Aurel Stein, an archeologist who spent more than a quarter of a century uncovering remnants of ancient civilizations that had long lain abandoned in the desert vastness.

The high moment of Stein's career came one day in 1907 near the city of Jiayuguan, where he spotted in the distance the ruins of a watchtower that "rose in a solid mass of brickwork, about fifteen feet square, to a height of some twenty-three feet." It was the westernmost extremity of the Great Wall of China, built nearly 2,000 years before and, in the dilapidation wrought by wind, sand and time, lost for centuries to human view.

Unlike the highland deserts of central Asia, much of northern Africa's gigantic Sahara is low-lying; its huge Qattara Depression is 436 feet below sea level at its lowest point. The difference in elevation accounts in part for a dramatic contrast in climate: Where the Gobi is consistently cold, the Sahara is a furnace by day and — because the desert surface reflects the sun's rays rather than absorbing and holding the heat — chilly by night, when temperatures drop by as much as 100° F.

The name Sahara — the Arabic word for "desert" — refers collectively to the several deserts that make up most of the northern third of Africa. About 3,000 miles wide and 1,000 miles deep, roughly equal in area to the entire United States, the Sahara is more than 3.5 million square miles of suffocating heat, almost constant wind and, outside the Nile Valley, fewer than 800 square miles of oases — those fertile areas where improbable eruptions of groundwater make agriculture possible. Only 20 per cent of the entire region is sand; the rest is rock, pebble and salt flats broken in places by mountains. In general, the Sahara is so bitterly inhospitable that fewer than

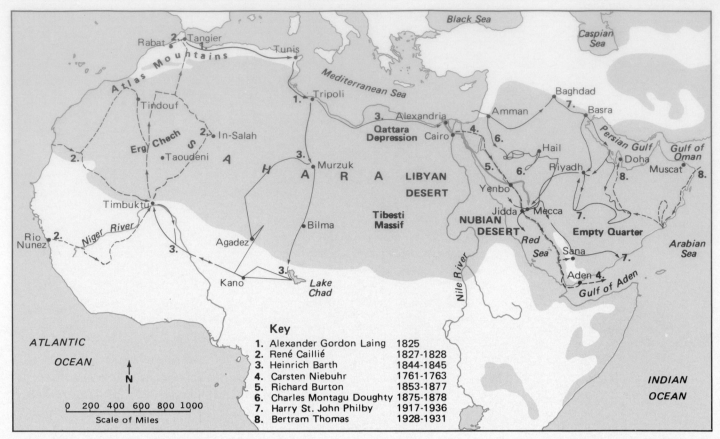

Key
1. Alexander Gordon Laing — 1825
2. René Caillié — 1827-1828
3. Heinrich Barth — 1844-1845
4. Carsten Niebuhr — 1761-1763
5. Richard Burton — 1853-1877
6. Charles Montagu Doughty — 1875-1878
7. Harry St. John Philby — 1917-1936
8. Bertram Thomas — 1928-1931

The 5,000 miles of desert from the Atlantic Ocean to the Arabian Sea is crisscrossed by the routes of 19th and 20th Century explorers, several of whom lost their lives while charting the region for the first time.

two million nomads are year-round inhabitants of its awesome reaches.

Because the trade routes carrying gold from Africa and spices from the Orient were controlled by Muslim caravaneers who demanded a large share of the profits, the Sahara was especially enticing to European merchants wishing to break the transport monopoly. They were, however, successfully fended off by a powerful Arab empire that for centuries ruled most of northern Africa, and not until the 19th Century, with the decline of the Islamic civilization, were Europeans able to penetrate to the interior of the Sahara.

Even then they were greeted with hostility. In 1825, Captain Alexander Gordon Laing, a Scotsman commissioned by the British government to travel south from Tripoli to the fabled city of Timbuktu, achieved his goal only to be strangled by his guide. Two years later, a young Frenchman named René Caillié, entranced by "the map of Africa, on which I could see nothing but blank areas," endured incredible hardships to reach Timbuktu; on one desperate stretch, the caravan to which he had attached himself went eight days without reaching water. And then came crushing disappointment. Unbeknownst to the Europeans, the great commercial and learning center of Timbuktu had all but disintegrated over the centuries. The so-called "Queen of the Desert," wrote Caillié, "consisted of nothing but badly built houses of clay, surrounded by yellowish-white shifting sands. The streets were monotonous and melancholic like the desert. No birds could be heard singing from the roof-tops."

For his honesty, Caillié paid dearly. Although he was hailed as a hero when he finally returned to France, reaction to his dismal description of the Sahara's gem city soon set in: Illusions die hard, and rumors were spread — and believed — that Caillié was a fraud who had invented his whole story. Embittered and disillusioned, Caillié became a recluse, and only 10 years after his return from Africa he died at the age of 39.

Yet in the early 1850s Caillié's description of Timbuktu was corroborated by another European who went there and subsequently declared that

Caillié had been "one of the most reliable explorers of Africa." That was praise indeed, since it came from Europe's greatest Saharan explorer — Heinrich Barth of Germany.

The son of a wealthy merchant often absent from home, Barth was an aloof child who became a young man widely disliked for his arrogance and general fussiness. Yet in the Sahara, where he found his only fulfillment, those disagreeable traits translated into self-reliance, confidence, fortitude and a meticulous devotion to the small details that can, in the desert, make the difference between death and survival.

Little regarded in his homeland, although he had earned a doctorate in philosophy at the University of Berlin, Barth eagerly grasped at an opportunity to join a British-sponsored expedition sent in 1849 to the Sahara south of Tripoli to suppress the slave traffic and to encourage more humane forms of commerce. During his long stay there, both of his European companions fell victim to the cruel hardships imposed by the desert. But Barth thrived. "Instead of feeling depressed at the death of my friends," he later wrote, "I felt that my forces were redoubled. There was a giant's strength in me!"

He was the quintessential loner, forever branching off from his caravan to investigate whatever attracted his interest. Unlike many other desert travelers, who preferred to move in the cool of night, Barth was willing to suffer the sweltering heat of daylight "to continue my exact observations."

A Sahara caravan nears Timbuktu, the fabled trade and cultural center on the Niger River in western Africa. This engraving illustrated the journal of the German explorer Heinrich Barth, who reached Timbuktu in 1853.

Exact they were. The Roman inscription on the ruins of a desert tower was found to be 32$\frac{7}{12}$ inches long and 15$\frac{10}{12}$ inches high; a well discovered at an elevation of 696 feet above sea level was five fathoms deep and the temperature of its water was a tepid 71.6° F. Much in the manner of a

mariner casting a log to gauge time and distance at sea, Barth used a chain to calculate the speed of his progress through the desert: He found that the normal rate was 2 miles per hour, but as supplies were consumed and the camels' burden was lightened, the pace picked up to 2½ miles per hour.

Barth had little inclination to marvel at the grandeur of desert scenery. Yet in his passion for specificity, he was careful to list the flora and fauna he saw along the way. Here, in their ceaseless struggle for survival, were the twisted trees and shrubs — the sidr, the ethel, the ghurdok; there were the venomous lizard, called bu-keshash, and little green birds, called asfir, which live off the vermin they peck from the feet of camels.

For five years, Heinrich Barth wandered in the desert wilderness; at length he was given up for dead, and his obituary was published. But in August 1855, Barth emerged hale and whole from the desert emptiness, bringing with him the notes and sketches from which derived his published work, a five-volume masterpiece of Saharan minutiae.

Back in Germany, he reverted to type: aloof, arrogant, fussy — and so unpopular that in 1857, despite his monumental accomplishments, he was rejected for membership in the Berlin Academy of Sciences.

Separated from the Sahara by no more than the narrow width of the Red Sea, the great Arabian Peninsula reaches 1,800 miles to the east and includes the world's most expansive vista of pure sand desert, with dunes rolling in endless succession to and beyond a horizon shimmering in the merciless heat of day. In the northern portion and along the Red Sea fringe of the 900,000-square-mile Arabian Desert, caravans of Arab traders, as well as pilgrims on their way to Mecca, have proceeded since ancient times. Thence too came the early European explorers: Carsten Niebuhr and Sir Richard Burton, sticking close to the Red Sea coast, penetrated as far south as Mecca, in some cases using Arab disguises to thwart the ban against non-Muslims in the region of Islam's holiest city; Charles Montagu Doughty and, later, T. E. Lawrence, along with others, probed much deeper into the desert interior.

Bertram Thomas is portrayed in Bedouin raiment after becoming the first Westerner to cross the 600-mile Empty Quarter of southern Arabia in 1930. Thomas found his Bedouin companions to be impulsively generous and protective of the weak — qualities he credited to the "hard school" of desert life.

Yet farther to the south yawned an enormous tract of sand that continued to defy the most daring of the Europeans. Lawrence of Arabia, for one, had insisted that it could only be surveyed by airplane. It was called the Rub'al Khali — the Empty Quarter — and, beyond a handful of nomadic tribes that were at constant war for possession of a few water holes, even most Arabs avoided the forbidding sea of endless sand.

It was here that a British civil servant named Bertram Thomas turned his face to the desert sun. Thomas was one of the many Englishmen who went to Arabia during World War I, when the region was at contest between Great Britain and Turkey. Unlike most, Thomas remained after the War, serving as vizier, or counselor, to the Sultan of Oman. His mind, however, was evidently not entirely on Arab affairs of state. "To have laboured in Arabia," he wrote, "is to have tasted inevitably of her seduction." And what allured him most was the Empty Quarter.

He left for the Rub'al Khali in December 1930, and his experience there was in large degree a study in sand. Approaching the immense dune country, Thomas saw "a vast ocean of billowing sands, here tilted into sudden frowning heights, and there falling to gentle valleys merciful for camels, though without a scrap of verdure in view. Dunes of all sizes, unsymmetrical in relation to one another, but with the same exquisite roundness of a girl's breasts, rise tier upon tier like a mighty mountain system."

Midway in the journey, at about 4:15 one afternoon, Thomas was startled by "a loud droning on a musical note" that emanated from a 100-foot-high sand cliff not far away. He had heard that the desert sometimes produced "singing sand," but this was more like "the siren of a moderate-sized steamship." Thomas searched for some sort of "funnel-shaped sand gorge that by some rushing wind action might account for so great a volume of noise." But he looked in vain, there was no such landform in the vicinity, and the only wind was a gentle breeze from the north. The phenomenon remained unexplained.

On the move for nine or 10 hours each day and stopping only for sleep or to allow the camels to forage at infrequent patches of scrubby brush, Thomas and his party passed beyond an area of rose-red sands and entered upon a region of fine white sand. It was, he wrote, "a scene of utter desolation, a hungry void and an abode of death to whoever should loiter there."

Just as he came to love the sand in its array of colors and forms, so he learned to fear it. At one point the party came upon what appeared to be a smooth salt plain but was instead a dry and powdery quicksand capable of swallowing the entire length of a six-fathom plumb line. Thomas learned too that the sand can be lethally treacherous even in the desert's rare havens. Pausing at a well, he was told by a tribesman that "four of my brothers lie in the bottom there. Two of them had descended to clean it out and were overwhelmed by slipping sand, and their companions, following to rescue them, were engulfed too. The well is a tomb. We have abandoned it."

Finally, almost two months after he had set out, Bertram Thomas emerged from the desert at the town of Doha on the Qatar peninsula on the Persian Gulf. He had covered more than 600 miles, and he had become the first Westerner to cross the Rub'al Khali. Thomas' feat was soon matched by two other intrepid British explorers, Harry St. John Philby in 1932 and Wilfred Thesiger in 1947; yet the Empty Quarter remains, with good reason, one of the loneliest places on earth.

From Francis Younghusband to René Caillié, from Bertram Thomas to Charles Montagu Doughty, men have been drawn into the arid wastelands by compulsions as various as the deserts themselves. In their wanderings they have pitted their skills, their stamina, their ingenuity, their courage and indeed their lives against one of the cruelest natural conditions existing upon the planet. And nowhere has the contest been waged more bitterly, at greater cost to life or with more satisfying triumph of the human spirit than on the huge island continent of the Southern Hemisphere, where winter comes in July and summer reaches its stifling peak in January.

"No country offers less assistance to first settlers," wrote Captain Arthur Phillip of the British Royal Navy long after he had first gazed on Australia's shores. Phillip had come to Australia in 1788 for an unpleasant purpose: He had been assigned to establish a colony for convicts. With the loss of its American colonies, England had been deprived of a place to send the overflow from its prisons—and nowhere seemed better suited as a substitute than the remote and empty continent Captain James Cook had claimed for Britain only 18 years before.

The site selected was Port Jackson—now the city of Sydney—on Aus-

tralia's southeastern coast. In the manner of such enterprises, tradesmen followed the convicts, farmers followed the tradesmen, and other communities sprang up to the extent that the coastal strip could no longer support the population. The government offered settlers large tracts of land in the interior, but the way was barred by the grim line of mountains belonging to the Great Dividing Range, which stretched the entire length of the eastern and southeastern coasts at distances of 20 to 200 miles from the sea.

Time after time the settlers attempted to breach the mountain barrier. Time after time they failed, and it was not until 1813 that a surveyor named George William Evans finally broke through to the plains that lay beyond. During the long and arduous efforts to traverse the Great Dividing Range, a remarkable fact had been noticed: The major mountain rivers ran not east to the sea but westward into the interior.

On the reasonable assumption that the rivers must empty into something, a pleasing theory was born: Somewhere far inland there must exist a great lake, the center of a region far more bountiful than anything the settlers had yet experienced.

That prospect was to dominate Australian exploration for the next 50 years. One after another the expeditions set out high with hope; one after another they met with shattering discouragement. Rivers became marshes and then disappeared entirely, evaporated by the fierce sun. The land became drier and drier, bleaker and bleaker. Yet on, over the years, they pushed. The dream of the inland lake gradually disintegrated. Exploration of the Australian Outback — as the interior was by now called — became, no matter what lay there, an end in itself.

The explorers finally succeeded not only in penetrating the heart of the Outback but in actually crossing the continent from south to north, by

34

Photographed during Philby's 1932 travels, Arab men remove from a desert well the rafters and animal skins placed by previous visitors to keep out sand. On learning of the remarkable sweetness of the water in this particular well, the explorer sent a sample for analysis; the results showed it contained camel urine.

Robert O'Hara Burke and William John Wills in 1861, and again by John McDouall Stuart in 1862. A decade later, 60-year-old Peter Warburton struck westward on camelback from Alice Springs, at the center of the continent. Ten tortuous months later he reached the seacoast—a full 1,000 miles north of his intended destination, Perth.

What the explorers found were endless square miles of sterile land, ravaged by time (geologically, Australia is probably the world's oldest continent) and climate; sweeping flats of baked mud, salt or flinty rocks; the infernos of the Simpson and Gibson Deserts; a ghost world in which twisted eucalyptus trees and clumps of thorny spinifex are bleached of their color by the pitiless sun.

For their achievements, the Australian explorers paid a terrible price. In 1831, Thomas Livingstone Mitchell turned back from the interior after two members of his party were killed by the region's aboriginal inhabitants. In 1840, John Baxter was also murdered by aborigines. In 1847, Edmund Kennedy led a 13-member expedition of which only three survived; the others, including Kennedy, died either of disease or at the hands of aborigines. In 1848, Friedrich Wilhelm Ludwig Leichhardt and six others headed into the Outback—and no trace of them was ever found. And in 1861, during the return journey from their successful south-to-north exploration, Burke and Wills perished of thirst and starvation.

Nowhere on earth have the arid lands made greater demands, and nowhere have humans set their will against the desert with greater resolution. The epic of the Australian explorers was the epic of man against nature—and in the experience of Charles Sturt lay the experience of them all.

An Army veteran of Waterloo, he had been sent to Australia as escort for a group of convicts. In New South Wales, while serving as secretary to

Foraging for brush to feed his camel, a member
of explorer Wilfred Thesiger's party pauses
on the slope of a majestic dune in Arabia's
Empty Quarter in 1946. Areas of less firmly
packed sand absorb more of the scant rainfall,
thus giving vegetation a chance to grow.

37

Governor Ralph Darling, Sturt became convinced of the reality of the great inland lake. In his own mind he elevated it to the status of the "new Australian Caspian Sea," and in 1828 he fared forth to find it. Since it did not exist, the effort was of course unsuccessful, but the glare of the sun had cost him a temporary loss of vision, and it was 10 years before his eyesight and general health recovered to the point that he could try again. Finally, in 1844, Sturt once more moved into the Outback, this time at the head of an expedition of 15 men, eleven horses, 30 bullocks and 200 sheep intended for food along the way.

Heading northwestward, the party progressed nicely at first, and a jubilant Sturt wrote, "We seem on the high road to success." But by the end of 1844, a summer dry even by Outback standards had set in; fearing to venture farther into the parched, cracked land, Sturt and his companions stopped at a water hole about 600 miles from Sydney — and there they remained for six months of agony.

Soaring as high as 119° F. in the shade, with a mean temperature of more than 100° F., the heat was beyond belief. Recalling its torments, Sturt later wrote: "Under its effects every screw in our boxes had been drawn, and the horn handles of our instruments, as well as our combs, were split into fine laminae. The lead dropped out of our pencils. Our hair, as well as the wool on the sheep, ceased to grow, and our nails had become as brittle as glass. We found it difficult to write or draw, so rapidly did the fluid dry in our pens and brushes."

The nights were almost as bad as the days. "The dazzling brightness of the moon was one of the most distressing things we had to endure," Sturt wrote. "It was impossible indeed to shut out its light whichever way one turned, and its irritating effects were remarkable."

As winter approached, the weather turned cooler, and Sturt, now suffering from scurvy, again moved to the northwest. The party came onto a vast

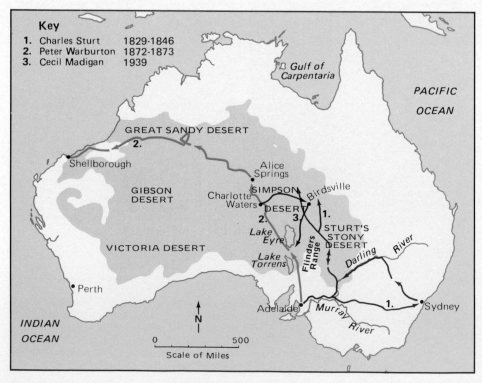

The forbidding deserts of Australia were not fully plumbed until relatively recent times, as the map at left shows. Indeed, the Simpson Desert at the heart of the continent defied mapping until an expedition headed by Cecil Madigan explored it on camelback in 1939.

plain covered by sharp, flintlike stones. Sturt named it the Stony Desert — and pushed beyond, moving into the dreadful expanse of what is now called the Simpson Desert, 43,500 square miles of uncharted scrub and dunes.

Finally the explorers could go no farther. Forced to turn back, they tried to avoid the Stony Desert but failed. "Coming suddenly on it, I almost lost my breath," Sturt wrote. "If anything it looked more forbidding than before. Herbless and treeless, it filled more than half the horizon."

By December the Australian summer had returned, and with it the intolerable heat. Sturt wrote, "The hot wind filled the air with an impalpable dust, through which the sun looked blood-red. So heated was the ground that our matches falling on it ignited."

On the evening of January 19, 1846, Charles Sturt and his companions stumbled into Adelaide after an absence of 18 months and a nightmarish journey of 3,000 miles. When, at midnight, Sturt opened the door to his home, his wife fainted. She had long since given him up for dead.

Today, a desert exploration such as Sturt's would be unthinkable — and unnecessary. After World War I, technology brought modern science to the deserts of the earth. For example, in the Gobi and its neighboring deserts, the wide-ranging expeditions during the 1920s of American naturalist Roy Chapman Andrews were made possible by a fleet of rugged automobiles (*pages 40-41*). Not only could Andrews quickly cover expanses that would previously have taken years to traverse, but he could carry with him far more scientific equipment.

Yet no matter how dedicated an explorer of the arid lands may be in his scientific pursuits, there must always lurk in heart and in mind a lingering sense of the desert's mystique. "We ourselves," wrote Andrews, "are the trail-breakers of motor transportation. Instead of pride at the thought, I reflected sadly that we were violating the sanctity of the desert." Ω

Members of Charles Sturt's 1844 expedition — the first to explore central Australia — survey the sandy hills near Lake Torrens in this contemporary watercolor. Sturt found the desert to be "one of the most gloomy regions man has ever traversed."

OPENING A LOST WORLD TO SCIENCE

Inspired by a theory that central Asia had once been a focal point in the evolution of the earth's mammals, American zoologist Roy Chapman Andrews in the 1920s led a meticulously planned and lavishly equipped assault on the secrets of the vast Gobi desert of Mongolia and China. His plans were greeted with skepticism, even ridicule. He might as well, said one critic, "search for fossils in the Pacific Ocean."

Undeterred, Andrews trekked into the Gobi five times between 1922 and 1930, at the head of expeditions that were remarkable even at the outset for two innovations, one scientific, the other practical. Andrews insisted on multidisciplinary teams, including experts on geology, topography and botany as well as the various zoological fields. He also introduced automobiles to Asian desert exploration. Because of the fierce winter weather, traveling had to be done between April and October, and the speedy cars allowed side trips and additional investigations that would not otherwise have been possible.

Andrews' second-in-command greeted the first of many discoveries that proved the critics wrong with a terse understatement: "The stuff is here." The prehistoric Gobi had indeed been home to a surprising diversity of animal life, including such mammals as mastodons, rhinoceroses and boars. In fact, the region was one of the richest fossil fields on earth. "In spite of the pessimistic predictions before our start," Andrews proudly proclaimed, "we had opened a new world to science."

A roaring convoy of Dodge cars carries members of the 1928 Andrews expedition across the Gobi desert in eastern Mongolia. The cars traveled at 10 times the speed of camel caravans and, having proved their desertworthiness, were eventually sold at a profit to local fur and wool dealers.

Relaxing at day's end in their camp at Chimney Butte, members of the 1928 Andrews expedition gather to listen to their Victrola. In defense of such amenities, Andrews (*fourth from left*) said, "I do not believe in hardships. They are a great nuisance."

The entire expedition of 1928 assembles for a portrait. The camel caravan would depart for prearranged campsites weeks ahead of the cars, each beast carrying 400 pounds of supplies — notably the gasoline required by the vehicles.

A major triumph of the Andrews expeditions was the discovery in 1923 of the first fossilized dinosaur eggs known to science. The nine-inch-long specimens shown here, uncovered during the 1925 expedition, justified the scientists' slogan: Bigger and better eggs.

The skull of a small, parrot-beaked dinosaur lies partially excavated from where it fell more than 70 million years ago. Discovered in 1922, the reptile was later named *Protoceratops andrewsi* in honor of the expedition's leader.

Chief paleontologist Walter Granger
(foreground), a member of all five Andrews
expeditions, gently lifts a nest of dinosaur eggs
from their lofty perch near the edge of the
Flaming Cliffs at Shabarakh Usu. The dramatic
outcrop, named for the brilliant hue it acquired
at sunrise and sunset, was the site of many of
the expedition's most important fossil finds.

Buried more than a thousand years earlier, a skeleton lies face down in its Gobi grave. Once clothed in a fiber robe decorated with beads made of clam shells, the individual was a member of a tribe that occupied the region before the advent of the Mongols.

With practiced ease, a Mongol photographed in the featureless Tsagan Nor basin of western Mongolia during the 1922 Andrews expedition singlehandedly erects a yurt to shelter his family. Ideally suited to the Mongol's nomadic life, the lightweight but sturdy portable structure has a thick felt covering to keep its occupants warm.

Archeologist Nels Nelson sorts through a portion of an enormous collection of artifacts assembled during the 1925 expedition. Most of the objects were discovered near the bed of a long-vanished river where a people christened Dune Dwellers by the Andrews group flourished for thousands of years.

Expedition leader Andrews edges out onto a precarious perch over what he described as "a wild chaos of ravines and canyons and gigantic chasms" to collect a young kite from its nest. During his five expeditions to the Gobi, zoologist Andrews collected more than 20,000 such specimens for study and display in museums around the world.

STUDIES IN DEPRIVATION

Alone in the airless, lifeless void of its orbit 140 miles above the earth in November of 1981, the United States space shuttle *Columbia* trained a special instrument on the surface of its home planet. The target was an anomalous brown patch on the predominantly water-covered blue orb of earth. The object of *Columbia's* attention was, in fact, a region only slightly less hostile to living things than space itself—North Africa's Western Desert.

The Western Desert is a place of terrifying barrenness that stretches from southern Egypt into the Sudan. It is part of the vast Sahara, one of the driest areas on the face of the earth, where decades may pass without a drop of rain. Hot winds have shaped the fine, loose sand into a landscape that is almost completely flat except for an occasional gently contoured sand hill or an outcropping of bedrock. One portion of the Western Desert is deeply mantled with sand; it is called the Selima sand sheet.

Columbia's experiment, little noticed amid the excitement of the space shuttle's second successful flight, involved a recently developed imaging radar system. The device beamed radar waves at the sand sheet and recorded the echoes for later processing into high-resolution images of the surface.

One of the agencies that received the data, and was especially interested in what it would reveal, was the United States Geological Survey, whose scientists had made a number of expeditions to the Western Desert to study desert land forms and processes. One of those scientists, Carol Breed, was the first to examine the radar images. Breed knew from firsthand experience that the Selima sand sheet was a flat and featureless expanse, and she was thus utterly unprepared for her first sight of the image. She experienced a moment of profound shock: "My God," she remembers thinking, "Where is the sand sheet?"

In more humid regions, moisture in the soil reflects and absorbs radar waves; but the sand of the Selima sand sheet, it turned out, is so dry that it is almost completely transparent to radar, and the signals returning to *Columbia's* instruments had been reflected by the bedrock under as much as 15 feet of sand. Instead of a monotonous plain of sand, the radar had recorded a startling picture of an ancient, rocky landscape of hills and valleys incised by large river channels. Some of the valleys were as wide as that of Egypt's Nile and were linked with a maze of smaller riverbeds, some no more than a few hundred feet wide. Here was the visible tracing of a rainy and far more hospitable past.

The recently unraveled history of the Selima sand sheet and the rest of the Sahara illustrates a truth about all the world's deserts: They are constantly

Moist winds from the south pile great banks of clouds against the Himalayas in Nepal. But the 26,000-foot peaks prevent the moisture from passing, and the resulting rain shadow accounts for the parched plain in the foreground.

moving across the face of the planet. Scientists estimate that the river systems etched in the Selima rock are at most 35 million years old. In fact, all the lands that are now deserts were at some time in the earth's 4.5-billion-year history covered by water, ice or lush vegetation. The evolution of arid lands, which continues today, is the result of titanic changes in the earth's crustal anatomy — changes that have been understood only since the maturing of the theory of plate tectonics in the late 1960s.

The theory holds that the earth's crust is divided into a number of huge plates, each constantly being augmented by volcanic activity at the mid-ocean ridges; that the plates creep outward from the ridges across the face of the planet at a rate of a few inches per year until eventually they are thrust down into the hot depths to be remelted; and that they carry with them the continents, which are continually torn apart and thrust together in new configurations by the inexorable movements. This notion of a dynamic earth helps explain such diverse geological phenomena as earthquakes, volcanoes, continental drift and mountain building. And it permits a reconstruction of the planet's history that helps diagnose the affliction of aridity.

About 140 million years ago, the supercontinent Pangaea, which then comprised all the land on earth, began to break apart. At that time shallow seas hundreds of millions of years old covered much of what is now dry land; 100 million years ago the seas began a retreat that over time removed a major source of moisture from the interiors of large continents. The South Atlantic, on the other hand, slowly spread open, with South America and Africa becoming completely separated only 85 million years ago. Meanwhile a land mass comprising what is now Antarctica, Australia and India broke away, then splintered into its constituent parts; Antarctica wheeling to the south, Australia to the east and India northward, to collide with Asia about 40 million years ago and begin the building of the Himalayas.

Just 35 million years ago, the Arabian Peninsula was still joined to the African continent, and a broad sea connected the embryonic Mediterranean to the Indian Ocean. Prevailing easterly winds gathered moisture from this now-vanished seaway and provided the lands of eastern Africa with abundant rainfall. But about 24 million years ago, a great rift began to develop between Africa and Arabia, slowly forming the Red Sea between them. In addition, the massive shifting raised mountains high enough to interfere with the flow of the rain-bearing easterlies and edged Africa and Arabia northward, narrowing the seaway and reducing the supply of moisture. By about 2.5 million years ago, the area's rivers and streams had dried up, and desert winds were beginning to pile sand in the valleys.

Since that time, the region has had perhaps 20 brief climatic reprieves. During the most recent one, which occurred about 10,000 years ago, the desert became a grassy savanna with a profusion of small streams and lakes and an abundance of animals that are found today only far to the south. But the dominant pattern of increasing aridity reasserted itself, the grasses and trees died, the soil turned to sand, and for the past 5,000 years this corner of Egypt and the Sudan has been a sand-blown desert with only a few scattered oases to sustain passing caravans.

There is a marvelous symmetry to the present-day distribution of the world's deserts (map, pages 80-81). Most of them lie in two globe-circling bands, one centered on the Tropic of Cancer at lat. 23°27' N. and the other

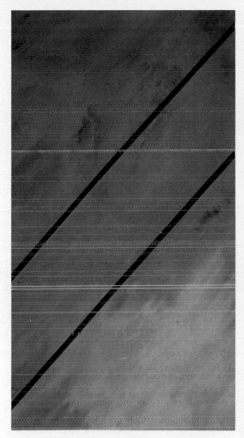

Dramatic evidence that the eastern Sahara was not always a desert is provided by a radar image taken from the space shuttle *Columbia* in 1981. A conventional satellite photograph *(above)* shows an almost featureless landscape, but *Columbia's* radar probed beneath the dry surface *(right)* to expose a complex of ancient river valleys under the sand.

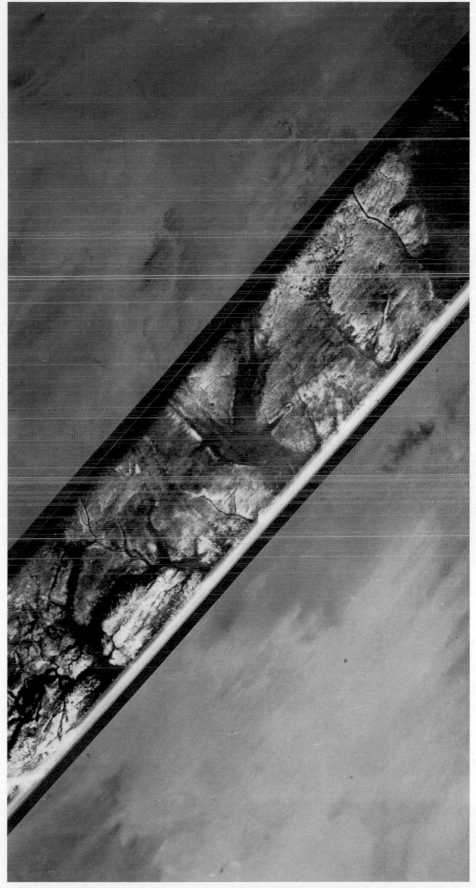

on the Tropic of Capricorn at lat. 23°27′ S. In the Old World subtropics of the Northern Hemisphere, a series of deserts stretches from the western coast of North Africa through the Arabian Peninsula and Iran into India. The New World counterparts are the deserts of the American Southwest and Mexico. Along the Tropic of Capricorn south of the Equator are found South Africa's Kalahari, the Monte desert of Argentina and the central deserts of Australia.

These deserts are placed and held along the edges of the tropics principally by the relatively constant patterns of circulation of the earth's atmosphere. Air is a fluid, and while the details of its behavior are complex in the extreme, on a large scale its movements can be described in relatively simple terms. The atmosphere operates as a kind of heat machine, kept in continuous motion by solar energy. When the sun's rays approach the earth, most pass through the atmosphere and are absorbed by land and water, then reradiated into the air as heat. The greatest proportion of the solar radiation reaching the earth is absorbed in the tropics, where the sun stands almost vertically overhead summer and winter; in other parts of the earth the arriving radiation strikes at a more oblique angle and consequently with less warming effect.

As the tropical air warms, it expands, becomes lighter than the surrounding air, and rises, carrying with it huge quantities of water vapor from the warm ocean surface. As the moist air rises it cools, until at an altitude of five to six miles it loses its buoyancy and begins to spread laterally, northward and southward. The cooling reduces the air's capacity to hold water vapor, and moisture begins to condense and fall in the deluges that are characteristic of tropical climates.

Further cooled and stripped of its water content, the increasingly heavy air sinks as it travels toward the Poles and is compressed by the continuing flow of sinking air. This compression causes the air to warm again — for every 1,000 feet of descent, the air temperature increases by approximately 3.5° F. This warm, dry, high-pressure air mass presses down on the earth's surface at about lat. 30° N. and S., and then much of it flows back toward the Equator into the low-pressure void left by the rising tropical air. The deserts of the subtropics are located where the parched high-pressure air descends; even the stretches of ocean that lie in the path of this desiccated air have little rainfall.

This system of air circulation generated by the tropical sun is called a Hadley cell, after George Hadley, the 18th Century British scientist who first described it. Two Hadley cells bracket the Equator, two similar systems of vertical circulation rotate over the temperate latitudes and one cell caps each Pole. Throughout most of geological time the pattern of the atmosphere's global circulation has been relatively constant and is expected to remain so for the life of the planet. Yet deserts are anything but constant, as the *Columbia* radar images of the Selima sand sheet's long-gone riverine systems clearly showed. The other factors that are at work were not understood until the development of the plate tectonics theory and the emergence of a clear picture of how the continents had reshaped themselves in the past.

That a mighty range of mountains can deny rainfall to the land in its lee has long been obvious; what plate tectonics helped explain was how even enormous ramparts of granite — the very symbols of permanence to awed humans—can be ever-changing, transient features. Great mountain ranges have been worn away literally to nothing, while the Himalayas, on the other hand, are relatively young mountains that are still being thrust upward by the grinding collision of India with Asia. The cataclysmic 1980 eruption of Mount St. Helens in the northwestern United States was but a dim echo of the volcanic frenzy that built the western mountain ranges of the Americas.

When wet air masses encounter high mountains, the air is deflected upward and grows cooler, and its moisture content condenses and falls as precipitation on the windward side of the mountain range. The aridity of the air that reaches the leeward side contributes to the formation of what is called a rain-shadow, or relief, desert. In the United States, the Cascade Range and the Sierra Nevada—created by the eruption of rock melted in the subduction of the Pacific Plate beneath the western edge of the North American Plate—make up a formidable north-south barrier rising to heights of 14,000 feet between the Pacific Ocean and the interior. The westerly winds dump as much as 100 inches of rain a year on the seaward slope of the mountains, enough to support lush rain forests in parts of the Pacific Northwest. Just to the east of the mountains lies the rain-shadow desert of the Great Basin, which stretches from the state of Washington south to include virtually all of Nevada and Utah. In eastern Australia, the Great Dividing Range deprives the interior of the moisture the prevailing easterly winds pick up over the Pacific. The southern U.S.S.R. and Afghanistan are arid because of mountains that block rain clouds coming from the Indian Ocean.

The effect of the rain shadow is heightened when it combines with another geographical factor—the sheer size of a continental land mass. The largest of the Asian deserts, China's Taklamakan and the Gobi of China and Mongolia, are dry in part because of mountain barriers but also because of their great distance from an ocean. This distance was increased dramatically 40 million years ago when tectonic movement added the expanse of the Indian subcontinent to that of Asia. By the time westerly winds reach central Asia, they have blown thousands of miles over land and gradually have lost virtually all of their moisture.

Surprisingly, some of the driest places on earth lie within sight of mighty oceans; deserts have developed along the western coastlines of three continents because the adjacent ocean water is cold. In each case, prevailing winds blowing generally along the coastline tend, because of an effect of the earth's rotation, to push surface currents seaward at right angles to the wind. Since there is no surface water upcurrent to replace the water being driven out to sea, very cold water is drawn upward from near the ocean floor. This vertical movement, known as upwelling, is reminiscent of the air circulation found in a Hadley cell. Air masses crossing these stretches of frigid water are chilled and their capacity to hold water vapor is consequently diminished. The condensing moisture either falls as rain at sea or forms dense banks of fog along the coast, leaving little or none to fall on the thirsty land.

Peru's Atacama Desert lies just east of an area of upwelling that has gained worldwide notoriety as a factor in the meteorological phenomenon known as El Niño. Every few years, usually around Christmas (hence the name, which is a colloquial Spanish term for the Christ child), the pattern

Fog casts a spectacular — but suspended and thus unavailing — shroud of moisture over the dunes of the Atacama, Peru's coastal desert. The winds off the

Pacific lose most of their rain-making capacity while crossing the frigid water offshore, leaving the Atacama one of the driest places on earth.

of the trade winds changes and warm water surges shoreward along the coast of South America. The obliteration of the upwelling area causes destructive rains, catastrophic alterations in the food supply for fish and birds and dramatic changes in global weather patterns. In normal years, however, the fog blowing in from the sea and occasional winter drizzles yield no more than 4/100 of an inch of precipitation for the Atacama in a year, and some spots go for years without a trace of rain.

Some of the most intriguing deserts in the world are exceptions to the general rule that arid lands are features of the subtropics. Forbidding cold deserts such as China's Taklamakan and the Turkestan deserts of the Soviet Union are located at considerable distances poleward from lat. 30° N., and the polar cells create their own set of desert conditions. Relatively warm air at 60° N. and S. — the latitude of Anchorage, Alaska, in the Northern Hemisphere and of open ocean below Tierra del Fuego in the South — rises and flows toward the Poles. As this air cools, it gives up its scant moisture as rain or, more often, snow, then sinks and moves outward to complete the circular flow.

Although there is abundant water in the form of thick ice sheets, glaciers and snow covering almost all the lands in the polar regions, by the standard of annual precipitation these areas qualify as deserts; even in the snowiest of years they receive no more than three or four inches — the equivalent of less than half an inch of rain. And scattered here and there within the Arctic and Antarctic Circles are barrens — ice-free expanses of rock or of gritty sediment deposited by glaciers where the scant snowfalls are quickly swept away by fierce polar winds. Parts of northern Greenland, the north slope of Alaska, some of the Canadian islands and one section of Antarctica harbor such patches of desert.

The brutal cold that characterizes the polar barrens (winter temperatures regularly drop to −70° F., and the warmest summer readings are no more than 25° F.) produces a geological feature not found in any other desert — permafrost, or permanently frozen ground. In the coldest of the polar deserts, permafrost extends from the surface to depths of more than 1,200 feet. Where there is an annual cycle of freezing and thawing, the permafrost is overlain by a so-called active layer. When this layer thaws in summer, the liquid water cannot trickle down but is instead trapped near the surface by the impenetrable barrier of the permafrost. As a result, it accumulates in puddles and shallow pools, creating the paradoxical condition of a waterlogged desert. One geologist has remarked, somewhat facetiously, that a person in a polar desert is more in danger of drowning than of dying of thirst when the active layer thaws.

Under the repeated assault of freezing and thawing, the active layer develops distinctive patterns. One is the result of a process called solifluction — the slow downhill creep of waterlogged soil in shallow, scalloped waves whose appearance is reminiscent of the wrinkled hide of an old elephant. The permafrost below provides a hard and fairly smooth surface over which the soil of the active layer can slide. The active layer can also be incised into polygons that bear a striking resemblance to the geometric forms that develop on playas, the seasonally dry bottoms of desert lakes. Whereas desiccation cleaves the surface of a hot-desert playa, in a polar desert it is cold that causes the soil to contract and crack. During the summer, water from melting snow trickles into the crevices, which can extend down into the permafrost. When the water freezes again, it forms an ice wedge that in the deeper, colder part of the crack persists throughout the year. In the driest of polar deserts, wind-blown sand rather than water may filter into the cracks.

The presence of desert sand in a frozen wasteland startled the British explorer Robert Scott when he discovered the first of the Antarctic desert areas — called dry, or ice-free, valleys. While crossing the continent's ice-covered highlands during an early attempt to reach the South Pole in 1903, Scott encountered extremely bitter cold and severe weather, despite the fact that it was December — summertime in the Southern Hemisphere. Following the course of a glacier, he descended into a valley to seek protection from the weather and found, to his astonishment, that the glacier did not spread out and smother the valley as usual, but dwindled away, then disappeared altogether.

Scott recorded his surprise in his journal: "Below lay the sandy stretches and confused boulder heaps of the valley floor. There was not a vestige of ice or snow to be seen, and as we ran the comparatively warm sand through our fingers, it seemed almost impossible that we could be within a hundred miles of the terrible conditions we had experienced. I cannot but think that this valley is a very wonderful place. We have seen today all the indications of colossal ice action and considerable water action, and yet neither of these agents is now at work. It is worthy of record, too, that we have seen no living thing, not even a moss or a lichen."

The valley into which Scott stumbled was one of three that run side by side in Victoria Land between McMurdo Sound and the Transantarctic Mountains. Snowfall is scant in the dry valleys, ranging to a maximum of four inches a year, and most of it is quickly swept away by ferocious winds of up to 120 miles per hour. What is more, the dark-colored rock and gravel exposed in the valleys absorbs so much heat from the sun that summer snows are quickly melted. The 15,000-foot-high mountains provide a barrier against the ice sheet that covers nearly all of eastern Antarctica; although here and there great outlet glaciers creep through gaps in the ranges to the sea, the ice-free valleys are, by geological happenstance, completely shielded. During the summer months, melting glacial ice feeds small lakes, many of which are laden with salts dissolved out of the surrounding rock. Though Scott was unable to detect any evidence of life, biologists have since discovered blue-green algae flourishing on the bottom of one of these lakes. The only other meager signs of life to be found are soil bacteria and a wingless fly — the largest land animal now living on the continent of Antarctica.

The boulders littering the floors of the dry valleys and the faces of the cliffs above are riddled with myriad hollows and tunnels that apparently are created by a combination of chemical disintegration of the rock and the abrasive effect of wind-blown sand and ice crystals. Some of these boulders, which are called tafoni, look like fantastic, rigid ruffles or petrified waves with tightly curving crests. Like their counterparts in the hot deserts, these rocks are being worked into sand by the chemical and mechanical effects of water, salt and wind. More sand is carried into the valleys by streams of meltwater flowing away from glaciers, with the result that there are actually sand dunes here in the midst of the frozen continent. They are small because

the sand supply is not great, but their forms are perfect polar images of the hot, high sand hills of the Sahara, the Taklamakan and all the other hyper-arid deserts that girdle the globe.

The harshness of the desert climate is reflected in the generally severe angularity of the landscape. Aridity helps to account for the hard-edged look. In more humid climates, water binds particles of sand, clay, silt and decayed organic matter into a cohesive mass of soil; tugged ever downward by gravity, the soil arranges itself into long, easy slopes that are further gentled by the presence of vegetation. But in the desert, flat basins of sand suddenly intersect steep, rocky slopes and cliffs. Sharp-edged boulders and barren expanses of rock predominate. Even the Sahara, which is famous for its expanses of sand dunes, is for the most part a desert of rock, with sand covering no more than 20 per cent of its total expanse.

The only permanent water supplies to be found in the desert are in oases, which are nourished by springs whose source is rain that may have fallen thousands of years ago, or thousands of miles away. The water is held in or

The constantly changing face of the desert is the result of relentless erosion. At left, sandstone in the Colorado Plateau, exposed to extremes of heat and cold, splinters away in thin chips that eventually will be ground into sand. In the Namib Desert *(above)*, the illusion of a "smoking" dune is created by an unflagging wind that lofts grains of sand over the crest of the dune, sculpting it to new shapes and moving it steadily leeward.

travels through an underground layer of porous rock called an aquifer *(page 63)*. In Africa, rain falling in the continent's highlands and equatorial region fills the pores of a layer of rock called Nubian sandstone that underlies much of the Sahara. The water percolates along slowly through the aquifer and is eventually brought to the surface at points where the underground layers have been folded or faulted.

The water of a fast-flowing aquifer is usually sweet and supports palm trees and other varieties of desert vegetation. But if the water moves slowly, it may leach out such large quantities of salts from the rock through which it passes that it is undrinkable and poisonous to most plants; only a few desert grasses and bushes have developed a tolerance for high levels of salts. Such plants flourish in the marshes that sometimes surround salty springs.

Narrow, ribbon-like oases form along rivers that rise outside desert areas. The longest of these rivers is the 4,000-mile-long Nile, which rises in the heart of equatorial Africa, where it is fed by heavy tropical rainfall, and flows through the eastern portion of the Sahara to the Mediterranean. A more modest example, the Barada river, originates in the mountains of

Spires of saguaro cacti rise from an unusual mantle of snow at Gates Pass, 3,172 feet above sea level in the Sonoran Desert outside Tucson, Arizona.

Temperatures in the Sonoran range from well below freezing in winter to a sizzling 120° F. at the height of summer.

Lebanon, descends to the Syrian Desert and dissipates in a vast marshy lake only 50 miles from its source. Short though it is, the Barada carries a sufficient volume of water to supply the one million people who live in the city of Damascus.

But just a few miles from such a water source the aridity that defines a desert — along with the intense heat that is characteristic of most arid places — is anathema to the human body. Any person caught in a hot desert without water and shelter has a life expectancy of only a few hours. In the summer, a gallon of water may be expended as sweat between sunrise and sunset. When the water thus lost is not replaced, the body gives up water stored in fat, tissue and the blood itself. The blood gradually thickens and flows more slowly, hence losing its effectiveness in cooling the vital organs. Moreover, the sweat glands fail from overwork and sunburn, with the result that high fever, delirium and circulatory failure lead to death. On an August day in the Sahara some years ago, a man and woman driving toward an oasis ran out of gas 10 miles from their destination. The man went for help, leaving the woman behind, and when he returned only five hours later she had died of thirst.

With very little water vapor present in the air to shield the ground from the full force of the sun's radiation or to help retain absorbed heat, most deserts alternate between hot days and cold nights. Air temperatures in hot deserts often exceed 100° F. The highest ever recorded — 134.6° F. — occurred in the town of Azizia in the Libyan Sahara on September 13, 1922. The record high in the United States was measured in California's Death Valley, part of the Mojave Desert, where the temperature reached 134° F. in July 1913. Temperatures on the ground, or within a few feet of it, are considerably higher — sometimes by as much as 50° F. (page 107).

However, the layer of desert surface that is heated is usually thin, since the dry materials of the desert floor are poor heat conductors. Thus the absorbed heat is quickly lost by reradiation at night and, because the humidity is generally very low, escapes upward without warming the air. As a result, very cold nights follow fiercely hot days. The lowest temperature ever recorded in the Sahara, not far from the location of the all-time high, is 20° F.

In a seeming paradox, water is the primary sculptor of the desert's face. The water can come from rain, snow or even dew. In Avdat in Israel's Negev, annual rainfall averages less than three inches a year, but on about 175 nights every year, the temperature dips low enough for dew to form, which adds another inch or so of precipitation to the annual total. However meager the supply, water is present and active at virtually every step in the long process by which mountains are fractured into great boulders, chiseled into a stony litter and, eventually, ground to sand and dust. The process begins as weathering — the disintegration of rock by chemical and mechanical means.

Given enough time, water seeping periodically into minute fissures in otherwise solid rock can work several kinds of destruction. Some of the rock's mineral components dissolve, causing ever-increasing structural weakness. Other minerals combine with water in compounds that are softer and more likely to crumble than the original materials. When water is absorbed by salts, they swell and exert pressure against the surrounding rock. If the rock becomes sufficiently hot, the water evaporates and the salts

The Unseen Origins of Oases

Oases are the miracles of the desert, providing water where no source seems to exist and making fertile small sections of otherwise lifeless land. The reality is that no desert is totally dry; somewhere underground, as the existence of oases attests, there is a continuous supply of flowing water. Its source is rain or snow that falls on higher ground nearby, or perhaps hundreds of miles away. Much of this precipitation runs off in rivers and streams, but a portion of it seeps downward until it encounters a layer of porous rock called an aquifer.

Gravity moves the water through the pores of the rock until, in one of a variety of ways, the water reaches the surface again. At a fracture point, or fault, in the aquifer *(bottom left)*, the seepage may be blocked by impervious rock that has been shifted into its path. The water then rises from the aquifer to the surface along the fault line. At other sites, erosion may lower the level of the desert enough, for instance on the floor of a basin *(bottom right)*, to expose the water-bearing stratum. In either case the result is a natural spring that transforms a bit of even the starkest landscape into a haven of green.

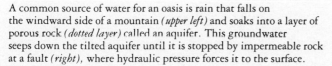
At the edge of China's Altyn Togh range, vegetation flourishes around a desert oasis fed from underground by rain that fell on the mountains miles away.

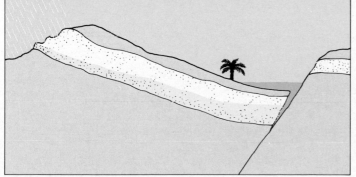

A common source of water for an oasis is rain that falls on the windward side of a mountain *(upper left)* and soaks into a layer of porous rock *(dotted layer)* called an aquifer. This groundwater seeps down the tilted aquifer until it is stopped by impermeable rock at a fault *(right)*, where hydraulic pressure forces it to the surface.

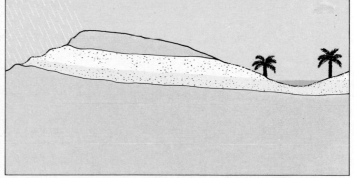

An oasis can also occur at a site where the erosive work of wind and sand, called deflation, has created a basin *(right)* lower than the elevation at which the rain fell *(upper left)*. Water in the saturated portion of the aquifer *(blue)* flows along its sloping course until it intersects with the desert surface.

shrink, only to expand when water once again trickles in. Over time, these small-scale changes in volume scar the rock with a complex pattern of tiny pits called honeycomb weathering, the best-known example of which can be seen on the face of the Sphinx.

Another kind of weathering has an awe-inspiring side effect that is the source of a number of desert legends. Alonzo Pond, a colleague of renowned explorer Roy Chapman Andrews, happened upon one example in the Sahara during the 1920s. While sitting on a high butte chatting with a tribal sheik one day, Pond asked whether there was anything interesting about the butte. "Yes," replied the sheik, "it sometimes shoots. It makes a loud noise like a gun." Intrigued, Pond asked when such a thing happened. "Oh, it generally shoots in the fall, when the summer heat is gone. It doesn't shoot every year, though. Only those years when something important happens."

Scientists have not established a connection between the shooting rocks and important events, but they have explained what causes the loud reports, and in the process contributes to the breaking down of rocks. Marked changes in temperature from day to night cause the rock surface to alternate continually between expansion and contraction. The repeated and irregular changes in volume create internal stresses within the rock surface that occasionally are relieved by a wrenching, explosive shift. Many desert travelers have reported hearing the sudden loud noises that result, but the event is rarely witnessed. Uwe George, a German naturalist who explored the Saha-

After a rainfall, hardy flowers blossom from cracks in the bed of a seasonal lake, or playa, in the Atacama Desert of northern Chile. As the playa dries out, the mud contracts into irregular slabs of clay divided by deep fissures.

ra on several expeditions during the 1960s and 1970s, wrote that he was once sitting near a three-foot-wide boulder on a hot summer day "when it suddenly gave a loud report, like a cannon being fired, and shattered into several pieces. Not long before, a brief but violent downpour of rain had cooled the heated stone so abruptly that its surface steamed."

Ice is another factor in the breaking down of rock, even in hot subtropical deserts. In parts of the Negev desert, the temperature falls below freezing on a dozen or more nights every year. And of course the cold deserts found at higher altitudes and latitudes experience much longer freezes; on the arid Colorado Plateau in southern Utah, for example, temperatures below −25° F. are common, and the Tarim River, which flows through China's Takla-makan Desert, is often completely covered by ice in winter. In such conditions, water trickling down into rock fissures freezes and expands, enlarging the cracks. When the ice melts, it trickles farther down into the extended fissures, where it eventually refreezes. Over the course of many winters, this repeated freezing and thawing will split a rock into pieces.

Dislodged chunks of rock exert their own mechanical force when they tumble down from a highland, striking and fracturing other rocks. The fragments often accumulate beneath rugged cliffs in steep-sided piles of rubble called taluses. As gravity and the runoff from torrential cloudbursts move the fragments downward and outward, weathering constantly replenishes the taluses from above. Where weathering is far advanced, there may be nothing left of the original highland except a mass of broken rock.

As a mountain is being reduced to rubble, its remains are further eroded by the action of wind and water. Collective runoff from intense desert cloudbursts commonly cuts straight-sided stream beds that are called arroyos in the American West and South America and wadis in North Africa and the Middle East. Because the desert supports little vegetation to impede its course, the water runs off in destructive torrents, making the arroyo a treacherous spot for the unwary. Experienced desert travelers never camp in an arroyo, since it is possible to be swept away by flood waters without any warning from rainfall, lightning or thunder. The only indication that a

One of the mysterious "moving rocks" of California's Death Valley leaves a telltale track across the bed of a dry lake. Though no one has ever seen them in motion, it is thought that the stones — some weighing several hundred pounds — are moved by strong winds occurring just after a rare rainfall, when the parched clay surface is temporarily wet and slippery.

distant, heavy downpour has launched a flash flood may be the roar of an approaching wall of water.

The torrent sweeps up vast quantities of debris, ranging in size up to large boulders. (Some geologists estimate that over the course of 60 million years, a layer of rock almost two miles thick has been removed by the forces of erosion from the Colorado Plateau.) As the speed of the water runoff decreases, the particles being carried along begin to settle out, the heaviest first. Channels leading down from a highland through hard rock to a plain or valley gradually decrease in depth as their angle of descent decreases, until they disappear altogether. Eventually the water spreads out in the shape of a fan, still carrying with it gravel, sand and other fine particles. Where these particles settle, they form a deposit called an alluvial fan. If several channels cut down through a highland close together, their alluvial fans overlap and finally merge into a bajada — an almost continuous deposit of sand and gravel skirting the foot of the highland. Another frequent sight in the desert is that of a mountain rising sharply from an encircling pediment — a base that resembles a bajada in shape but is made of bedrock instead of alluvial sand and gravel. Geologists are still debating whether meandering, intermittent streams or sheets of floodwater are responsible for shaping pediments.

Hot deserts, where the rate of evaporation far exceeds the rate of precipitation, are often dotted with playas, the dry lake beds that are temporarily covered by shallow water after a heavy rainfall. In winter, when evaporation is slower, a playa lake may persist for a month or so, but in summer the water usually disappears within a week. The evaporating water leaves a residue of fine clay, silt or salt that gradually builds up into a thick, level layer. The surface may slope no more than two or three feet per mile, making the playa the flattest landform on earth.

When a playa is surfaced with sodium chloride — ordinary salt — or calcium carbonate, it is hard enough to support cars, trucks and even landing spacecraft; Rogers Lake, a 65-square-mile playa in the Mojave Desert, has been used as a landing site for the space shuttle. But when a playa crusted over with clay or silt dries out, it becomes incised with deep cracks that intersect to form polygons, most of them five-sided. Some of the polygons are huge, measuring 250 feet across, and the fissures may be 15 feet deep and three feet wide.

Above the parched flats of a playa rise rocky islands, and peninsulas jut out from its terraced shores toward the center. These features were shaped thousands of years ago when permanent lakes occupied playa basins. During the ice ages that lasted from about 2.5 million years ago to about 10,000 years ago, glaciers in the Northern Hemisphere extended southward to the latitude of Cincinnati, Ohio. Cooler, rainier climates prevailed south of the glaciers in regions that are deserts today. The Southern Hemisphere experienced a similar cooling, and even at the Equator temperatures were a few degrees lower than they are today. Plentiful rain and a lower rate of evaporation combined to produce huge lakes. In Utah, for instance, Lake Bonneville covered 20,000 square miles to depths of 1,000 feet and more, and resembled present-day Lake Michigan. As the glaciers began their slow retreat northward at the end of the last Ice Age, a warmer, drier climate took hold in the Western United States, and Lake Bonneville gradually started shrinking. All that now remains of it is the 1,000-square-mile Great

Dubbed the "Great Ear" because of its shape, the large empty lake at the edge of the Taklamakan Desert in China's Sinkiang Province appears in this satellite photograph as a concentric series of shorelines. Evaporation during the dry season has left a pale residue of salt; moisture remaining in the bottom of the lake shows as darker brown.

In the aftermath of a cloudburst, a cataract of water roars over the terraced canyon walls of the Little Colorado River at Grand Falls, Arizona. Such

torrents, carrying all before them, sweep away enormous quantities of clay and rocky debris and carve dramatic features in the forlorn landscape.

Salt Lake which, should the climate become drier still, could disappear as a year-round feature.

The other major shaping force in the desert is wind. Its ability to erode the landscape is far greater in arid lands than in humid regions because there is scant vegetation to interrupt its flow and little water to bind small objects into a larger cohesive mass that would offer greater resistance. A strong desert wind can sandblast anything in its path. When Uwe George was driving across the Sahara with a party of explorers, they were caught in a fierce sandstorm whose winds reached hurricane strength. After two days, all the paint had been stripped from their vehicles, and the windshields were so pitted and abraded that the drivers could not see through them and had to knock the glass out entirely.

Sand-laden winds also carve rock in distinctive ways. Many deserts are peppered with ventifacts — rocks that have been shaped by centuries of wind abrasion and deflation. Outcroppings of sandstone and other soft sedimentary rocks — and even hard granite — are sculpted into yardangs, streamlined shapes that sometimes resemble the upside-down hulls of boats. Yardangs range in length from a few inches to a half mile and may rise 600 feet above the surrounding bedrock. Whatever their size, they are

Two streamlined sculptures called yardangs *(foreground)* face the Sahara's prevailing winds. Variations in the wind and in the hardness of different layers in these ancient chalk deposits account for the variety of sandblasted shapes.

70

always oriented in the direction of the prevailing wind and, when they appear in clusters, take on the appearance of a beached fleet of vessels. Sandblasting winds can also score deep grooves in the surface of soft rock. Satellite photographs of the Tibesti Massif, which stretches from Libya south into Chad, show a spectacular array of steep-sided grooves a quarter mile wide or more, some of them filled with sand dunes. The grooves, which extend across 35,000 square miles, are aligned with the prevailing northeasterly winds, which follow a slightly curving course around the mountain range.

Water and wind sometimes act together to form surfaces that are unique to the desert. One of these, called desert varnish, is a brownish or black coating that can give rocks of entirely different composition a similar veneer. The varnish, only a fraction of a millimeter thick, is thought to be layers of tiny particles of clay, and of oxides of iron and manganese, deposited on rock surfaces by wind and water from rain and dew. The oxides cement the clay particles and stain them a characteristic color. The exact time span required for a rock to develop an envelope of varnish has not been determined, but geologists think it may be as long as 20,000 years.

Another product of the teamwork of wind and water is called desert pavement, or *reg*, from an Arabic word meaning "little." It is a mosaic of pebble-sized fragments of extremely hard rock, such as flint, that is slow to disintegrate. Transported by water from highland areas to the desert floor, the pebbles of the pavement are mixed with sand, clay and silt. Over thousands of years, however, the wind picks up the small, lightweight particles and blows them away in a process called deflation. As the fine matrix is whisked away, the pebbles gradually settle together, interlocking as snugly as the stones of a cobbled street. In the Sahara and Egypt's Sinai desert, pavement extends over tens of thousands of square miles, monotonous, somber and lifeless, stretching as far as the eye can see with hardly any variations in texture, color or elevation.

The lightest of the materials that the wind removes from the desert floor are specks of clay and silt that measure no more than a few thousandths of an inch in diameter — so small that, even on a calm day, rising columns of hot air can carry them aloft to form a choking fog of dust. Strong winds sweep the particles to heights of 10,000 feet or more and carry them thousands of miles. Reddish dust often spreads northward from the Sahara over the Mediterranean, settling eventually on the rooftops of Paris and sometimes reaching even farther north, to Sweden. Semiarid lands often have thick, fertile layers of silt and clay that originated as dust blown from neighboring deserts. Called loess, these deposits can cover enormous areas. In northern China, loess that came from the Gobi desert of central Asia covers more than 300,000 square miles with a blanket of soil 700 feet thick.

The wind also moves the larger and heavier grains of sand that remain after the finer dust has been carried away, but in different ways. The average grain of sand is at least 25 times bigger than a particle of clay, and cannot be blown as easily, or as far, as dust. But given the right conditions and enough time, wind abetted by occasional running water will shepherd the sand into huge geologic depressions, named ergs by the Berber nomads of the Sahara. The largest erg in the world is the Arabian Peninsula's Rub'al Khali — the Empty Quarter — an uninterrupted expanse of sand measuring roughly 400 by 700 miles. In some areas the bedrock is covered by only a few inches of sand, while in others the wind piles abundant sand into dunes that rise as high as 700 feet and crest like the waves of a gritty sea.

Dune country is an ominous place for human beings. The very name of China's great sand desert, the Taklamakan, means "the place from which there is no return." But dune country is also a place of peculiarly riveting beauty. Ralph Bagnold, a British Army officer and scientist who was posted to Egypt's Western Desert in 1925, was captivated by the spectacle of the dunes. "The observer never fails to be amazed," he wrote later, "at a simplicity of form, an exactitude of repetition and a geometric order unknown in nature on a scale larger than that of crystalline structure. In places vast accumulations of sand weighing millions of tons move inexorably, in regular formation, over the surface of the country, growing, retaining their shape, even breeding, in a manner which, by its grotesque imitation of life, is vaguely disturbing to an imaginative mind. Elsewhere the dunes are cut to another pattern — lined up in parallel ranges, peak following peak in regular succession like the teeth of a monstrous saw for scores, even hundreds of miles, without a break and without a change of direction, over a landscape so flat that their formation cannot be influenced by any local geographical features."

Originally, Bagnold was interested not so much in the dunes themselves but in the remains of ancient Egyptian civilizations that he might find among them. He struck up a friendship with an officer who shared his interest in archeology, V. C. Holland, and soon the two of them began making short forays into the desert. They hoped, Bagnold wrote, that "in those unsurveyed expanses of sand and rock there might be something still to be discovered just a little farther out."

Their expeditions also offered them a different kind of excitement when they tried driving cars where less adventurous Englishmen said no cars could go. But Bagnold decided to test an advertising slogan then being used by the Ford Motor Company to emphasize the virtues of the Model T: "No hill too steep, no sand too deep." He found that, provided one did not take the slogan entirely at face value, the cars were useful. On one occasion, Bagnold reported, when driving up a particularly steep dune, "we rose as in a lift, smoothly, without vibration. We floated up and up on a yellow cloud. All the accustomed car movements had ceased; only the speedometer told us we were still moving fast. It was incredible."

During the decade he spent in Egypt, Bagnold became an avid student of the forms of sand dunes and the movements of the billions of sand grains that shape a dune. Upon his return to England, he prevailed upon the Imperial College of Science and Technology in London to let him use its laboratories and wind tunnels for a series of pioneering experiments on the interaction of sand and wind.

Bagnold found that the wind must reach a specific speed, determined by the size of the sand grains, before they begin to roll along the surface. For one type of sand he tested, the velocity was 11 miles per hour. When a rolling grain encounters a stationary one, the impact might knock the second grain forward like a billiard ball or might launch it into the air. A fast-moving grain that strikes a pebble or other large obstacle bounces into the air. But the flight of any one sand grain is usually short (even in the worst sandstorms the particles seldom reach heights of more than six feet) and when the particle lands it knocks other grains this way and that in what can

Desert pavement, a dense cover of rock fragments too large to be carried away by wind or water, coats a stretch of land in Arizona. Silt, sand and smaller pebbles have been removed by a gradual erosive process called deflation *(diagram, below)*. After many centuries, the remaining stones have settled into a coarse mosaic, resembling a street paved with cobblestones, that presents a shield against further erosion.

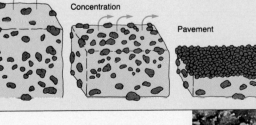

Deflation

Concentration

Pavement

73

BARCHANS
Barchans form graceful crescents across the prevailing winds of the Vizcaino Desert in Baja California, with their tips, or horns, pointed downwind. Barchans are the most mobile of all dunes; the ones shown here migrate as much as 50 feet per year.

STAR DUNES
Constantly shifting winds can assemble several sharp-crested ridges, each formed across the wind like a barchan, into a star-shaped dune such as the one shown below in the Namib Desert. Because the wind direction is so erratic, star dunes do not migrate.

SEIFS

Slender longitudinal dunes — called seifs, the Arab word for "sword," because of their blade-sharp crests — run parallel to the prevailing winds in this aerial photograph of Egypt's Western Desert. Where sand is abundant, dunes form along the wind instead of across it; minor variations in wind direction account for their sinuous course.

LINEAR DUNES

Ranks of parallel longitudinal dunes dotted with shrubs stretch to the horizon in Australia's Simpson Desert. These long, rounded dunes, which are sometimes 200 miles long, are formed when strong, steady winds cut deep troughs in the desert floor and pile the sand particles in symmetrical configurations on either side.

best be described as a splash. If the wind continues blowing, the air near the surface is soon filled with bouncing sand grains, while larger particles roll along the surface with the wind.

If a stream of moving sand encounters an obstruction such as a boulder, a plant or even a small irregularity in the bedrock underlying the sand, the air flow is disrupted. Immediately in front of the obstacle, and to an even greater degree just behind it, wind velocity drops. In these two pockets of slow-moving air, sand grains come to rest and begin to pile up. The accumulation behind the obstruction is at first the larger of the two, but they soon coalesce into a single mound, and the formation of a dune has begun. The shape it takes as it grows will be determined by the velocity and steadiness of the wind and sometimes by the amount of sand available.

Where the wind direction is generally uniform and the amount of sand moderate, the dune assumes a crescent shape and is called a barchan. The tips of the crescent point downwind and are lower than its center, where air flow is impeded most and sand consequently accumulates in larger quantities. Where sand is more plentiful, a steady wind often creates transverse dunes shaped like long, straight ocean waves with crests at right angles to the wind and gentle windward slopes. When the wind occasionally shifts direction by a few degrees, long ridges of sand form parallel to the general direction of the wind instead of across it. Called longitudinal dunes, these ridges may be 60 miles long and hundreds of feet high. If the sand is confined to a basin, and the wind periodically changes its direction radically, the resulting dunes will assume the shape of a starfish—the most complex form of all—and are called star dunes (page 74).

The windward side of a dune slopes at a smaller angle than the steeper lee side. When sand grains are pushed up the windward slope to the crest, they fall over the top, which maintains a sharply defined edge. Because sand grains are continuously reaching the crest and dropping over the edge, the windward side is constantly eroding. When the sand falling over onto the leeward side piles up into a slope greater than 35 degrees, a little avalanche occurs, restoring what is called the angle of repose—the maximum angle, formed by the intersection of the side of the dune with the ground, at which the sand grains can remain at rest. Because of these repeated little slippages of sand, the lee side is also called the slip face.

As long as the wind blows hard enough, grains of sand are constantly migrating up the windward slope, over the crest and down. The smaller grains move along faster than the coarser ones, which tend to accumulate on the windward face. That face is also more tightly packed because of the action of the wind and is often firm enough to support a truck. But, grain by grain, the substance of the dune is constantly rearranging itself and finally leaves the dune altogether. Yet all the while, as the individual grains migrate, the dune maintains its form, moving slowly downwind in the process. Dunes ordinarily slip along at a barely perceptible 10 feet a year, but where winds are strong and steady in direction, dunes can move as much as a foot a day.

The movement of sand also engenders another of the desert's eerie mysteries—that of the singing dunes. Marco Polo heard them during his journey through the sea of sand dunes in the Taklamakan Desert: "Often you fancy you are listening to the strains of many instruments," he wrote in his journal, "especially drums, and the clash of arms." When Richard

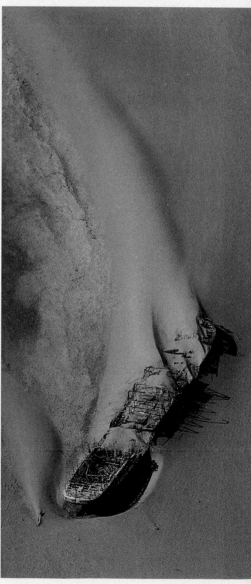

The striated longitudinal dunes of the Namib Desert end abruptly at the Kuiseb River in the photograph at right, taken from an orbiting satellite. Thwarted in their march northward, the dunes instead are expanding westward into the Atlantic Ocean, gradually changing the southwest coastline of Africa.

Sands drifting over an old shipwreck offer dramatic evidence of the Namib Desert's incursion into the sea. The German cargo ship *Eduard Bohlen,* which ran aground off the perilous African coast in 1912, is now more than half a mile inland.

A windswept field of sand and rock on the arid surface of Mars, photographed by an unmanned Viking landing vehicle *(foreground)*, strikingly resembles certain hyperarid deserts on Earth.

Trench traversed the dunes of the southern Sahara in 1974 he too heard strange sounds like the rumbling of drums. He was told by a cameleer in his caravan that this was "the laughter of Rul, the djinn of the dunes, who torments the traveler as he becomes disoriented by fear and thirst." Even scientists such as Ralph Bagnold attest to the existence of these peculiar sounds. "At times, especially on a still evening after a windy day," Bagnold reported, "the dunes emit, suddenly, spontaneously, and for many minutes, a low-pitched sound so penetrating that normal speech can be heard only with difficulty."

Bagnold could find no satisfactory explanation for this phenomenon, nor has anyone managed to discover the distinction between noisy and silent dunes. The British scholar H. S. Palmer studied a singing dune in the western Sinai that, according to local legend, long ago buried a Christian monastery. Now and again a wooden gong in the monastery is thought to begin sounding loudly enough to be heard through the depths of piled-up sand. Palmer examined the sand of this dune and found it to consist of uniformly large grains of quartz. When set in motion by any disturbance, such as the passage of an animal or the wind, the sand begins to move down the slope in thin waves, like oil over glass. As it moves, what Palmer described as a "deep vibratory moan" increases to a roar, then dies out as the sand rivulets come to rest. Palmer discovered that the higher the temperature of the surface sand, the louder the noise it produces. Yet why the

By studying the wind and erosion patterns of this bleak section of Egypt's Western Desert, geologists expect to learn more about the similarly arid and hostile landscape of Mars

sand of this dune sings while that of other apparently similar dunes does not remains a mystery.

Many such mysteries remain in the sere expanses of the world's deserts — questions about how these barren places came into being, what preceded them, and what they might become next. Pursuit of the complex secrets leads inevitably to even larger considerations — of the origins and eventual fate of the planet and its life forms.

Thus on June 19, 1976, five years before the space shuttle *Columbia* unveiled the floor of the Selima sand sheet, a squat, ungainly spacecraft went into a different orbit and trained its instruments on a hyperarid region with features identical in many respects to those of Egypt's Western Desert. For several days it photographed networks of river channels etched in bedrock, broad ranks of crescent-shaped barchans, and great grooves and yardangs carved by stinging, sand-laden winds. A month later the three-legged craft left its orbit and parachuted down to the hostile surface, where it recorded closeup views of wind-drifted sand, weathered basalt rocks and colossal sandstorms. There it remained, transmitting to eager geologists reams of information about the desert plains of Memnonia — on the planet Mars. Detailed comparisons of the data from the *Viking 1* Mars probe with information about the Sahara gathered by satellite, space shuttle and on foot continue to shed new light on the geological history and processes of the deserts of both planets. **Ω**

THE ENDLESS MIGRATIONS OF ARID LANDS

Just as it is the nature of nomads to wander the face of the desert, so it is the nature of deserts to wander the face of the globe. And like the nomad who never migrates aimlessly, the desert moves in predictable response to specific geologic and atmospheric conditions.

Four factors conspire to deny rainfall to arid lands. The most constant of these is the global circulation of the atmosphere itself, which maintains twin belts of dry, high-pressure air over the fringes of the tropics. Circulation patterns in the sea also make a contribution to aridity when cold coastal waters chill the air, reducing its moisture-carrying capacity. But even water-laden winds may not carry rain to a parched desert if it is in a so-called rain shadow created by a mountain range, or if the distance to the interior of a continent is too great.

By calculating the effects of these conditions, Christopher Scotese, a geologist at the University of Chicago, and Judith Parrish of the U.S. Geological Survey have charted the advance and retreat of arid lands across the ever-changing arrangement of the continents during the past 65 million years. They also have projected the appearance of an increasingly arid earth 100 million years hence.

Present-day Australia may be a harbinger of the future. As the continent has inched northward to its present position, centered on lat. 30° S., the combination of the overlying belt of high pressure, the rain shadow created by coastal mountains and the tremendous distance between the interior and the ocean has caused arid and semiarid lands to spread across two thirds of the continent.

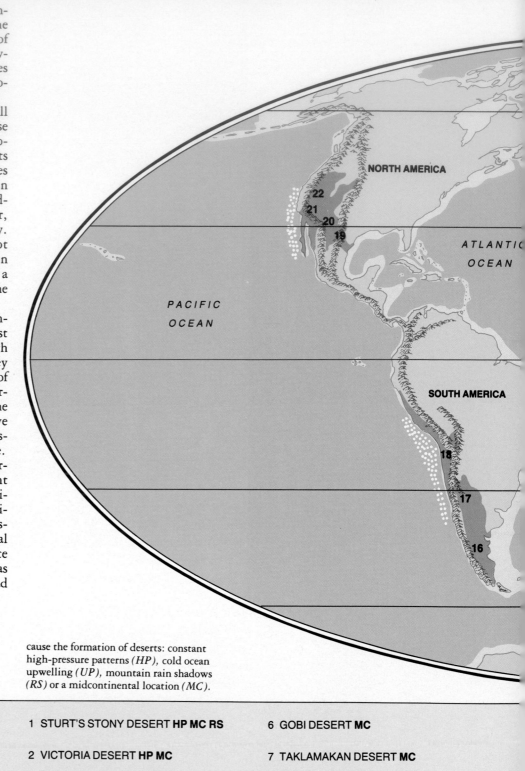

A map of the world's desert areas shows their tendency to cluster in the subtropics. As indicated on the chart below, each is the product of one or more of the four factors that cause the formation of deserts: constant high-pressure patterns (HP), cold ocean upwelling (UP), mountain rain shadows (RS) or a midcontinental location (MC).

HP=HIGH PRESSURE	1 STURT'S STONY DESERT **HP MC RS**	6 GOBI DESERT **MC**
MC=MIDCONTINENT	2 VICTORIA DESERT **HP MC**	7 TAKLAMAKAN DESERT **MC**
UP=UPWELLING	3 GIBSON DESERT **HP MC RS**	8 THAR DESERT **HP**
RS=RAIN SHADOW	4 SIMPSON DESERT **HP MC**	9 IRANIAN DESERT **HP MC RS**
	5 GREAT SANDY DESERT **HP MC**	10 TURKESTAN DESERT **MC**

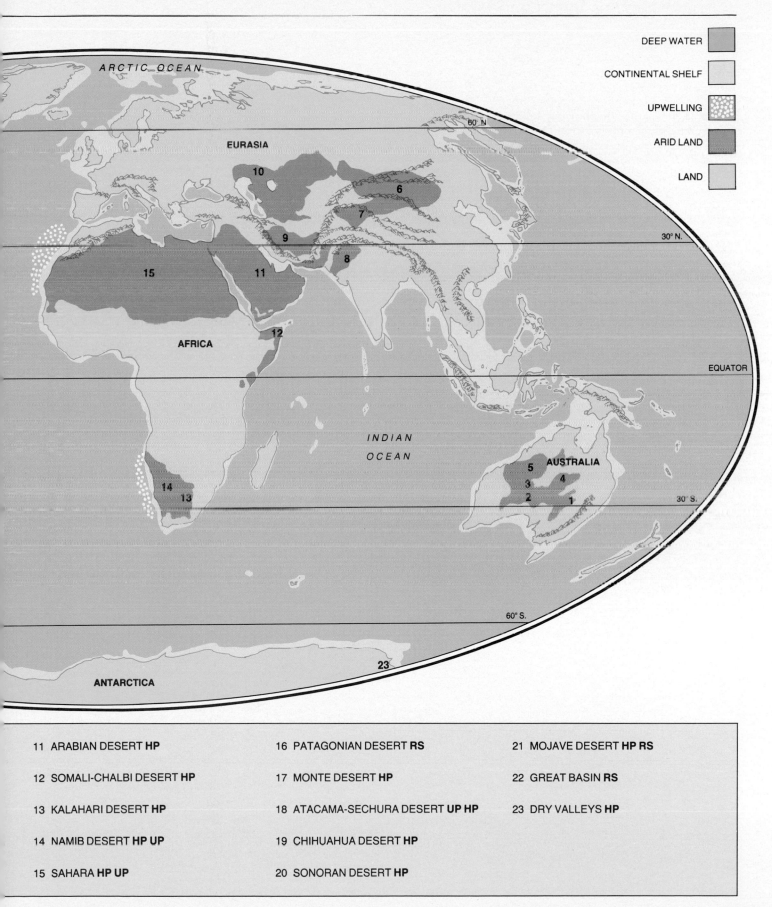

ARCTIC OCEAN

EURASIA

80° N

30° N.

AFRICA

EQUATOR

INDIAN

OCEAN

AUSTRALIA

30° S.

60° S.

ANTARCTICA

11 ARABIAN DESERT **HP**	16 PATAGONIAN DESERT **RS**	21 MOJAVE DESERT **HP RS**
12 SOMALI-CHALBI DESERT **HP**	17 MONTE DESERT **HP**	22 GREAT BASIN **RS**
13 KALAHARI DESERT **HP**	18 ATACAMA-SECHURA DESERT **UP HP**	23 DRY VALLEYS **HP**
14 NAMIB DESERT **HP UP**	19 CHIHUAHUA DESERT **HP**	
15 SAHARA **HP UP**	20 SONORAN DESERT **HP**	

A Conspiracy of Wind and Water

Sixty-five million years ago, great deserts had formed on several of the continents created by the breakup of the supercontinent Pangaea. These deserts were the work of the two most enduring causes of aridity: bands of atmospheric high pressure over the subtropics, and areas of cold-water upwelling at sea.

The high-pressure areas are created by two great rotating cells of air, driven by hot air rising from the Equator, that encircle the earth over the tropics. At the apex of the cells, the cooling air begins to sink and reaches the surface again at about lat. 30° in each hemisphere. Having given up most of its moisture as it cooled, the descending air is compressed and warmed by the pressure of the air above it. Consequently, its capacity for moisture is increased, so that little precipitation occurs, surface moisture evaporates quickly and any land in these latitudes is likely to be parched.

The effect can be compounded by cold upwelling along the west coasts of subtropic land masses, where prevailing winds push ocean-surface water seaward in such a way that it can only be replaced by very cold water from the depths. Moist air masses that might otherwise bring precipitation inland are chilled, and their moisture either falls offshore or wafts over the thirsty land as fog.

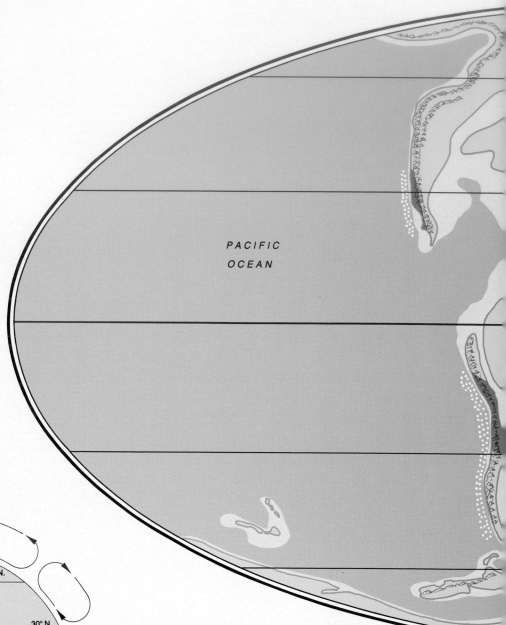

PACIFIC OCEAN

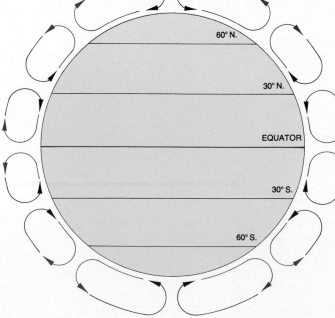

60° N.

30° N.

EQUATOR

30° S.

60° S.

A schematic diagram of the six endlessly cycling atmospheric cells that encircle the globe shows relatively warm air rising at the Equator and at lat. 60° N. and S. Dry, dense air descending in the subtropics of each hemisphere, and at each Pole, forms zones of high pressure that have a profound effect on the formation of major deserts.

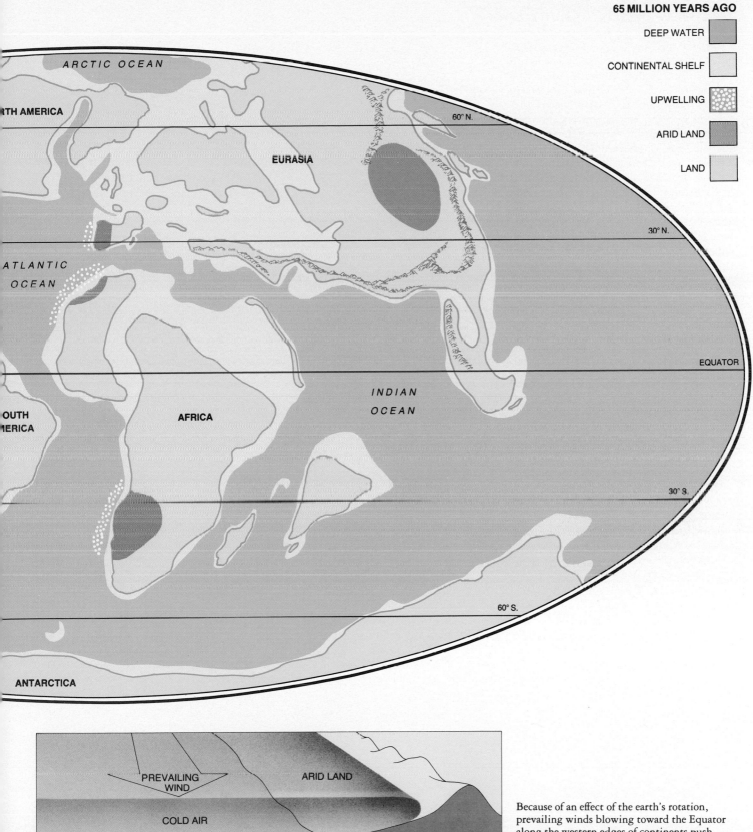

DEEP WATER

CONTINENTAL SHELF

UPWELLING

ARID LAND

LAND

ARCTIC OCEAN

NORTH AMERICA

EURASIA

60° N.

30° N.

ATLANTIC OCEAN

EQUATOR

INDIAN OCEAN

SOUTH AMERICA

AFRICA

30° S.

60° S.

ANTARCTICA

PREVAILING WIND

ARID LAND

COLD AIR

UPWELLING

Because of an effect of the earth's rotation, prevailing winds blowing toward the Equator along the western edges of continents push coastal waters seaward, at right angles to the wind. Underlying cold water drawn upward to replace the wind-driven current chills the sea surface and causes frequent fogs but prevents precipitation from reaching the coast.

The Rise of the Rain Shadows

About 45 million years ago, grinding collisions between sections of the earth's crust — the tectonic plates — were beginning to build titanic new mountain ranges that would dramatically alter the size and location of the world's deserts. Along the western coasts of the Americas, the floor of the Pacific Ocean was being thrust downward and melted under the continental plates in a process called subduction. The resulting pools of liquid rock erupted to the surface inland, building the Rocky Mountains, the Andes and the Sierra Madre.

The higher and more widespread the American mountain ranges became, the more completely did they shield continental interiors from moisture-laden air masses. As air moving inland was forced upward by the intervening peaks, it cooled, its moisture condensed out as precipitation, and only dry air continued inland. As a result, the Patagonian desert, east of the Andes near the southern tip of South America, and North America's Great Basin desert, bordered by the relatively young Sierra Nevada, achieved their great size. So pronounced is this effect, known as a rain shadow, that some scientists insist any region in the world could become a desert if blocked from water by mountains of sufficient size.

Mountain building increased the size of North Africa's desert area as well, but it did not affect Europe and Asia the same way. There, in fact, movement of the Eurasian Plate permitted the ocean to spread inland, filling a vast, low-lying midcontinent basin. Moisture from this great inland sea, carried by westerly winds across Asia, eliminated the enormous arid region that 20 million years earlier had covered more than two million square miles.

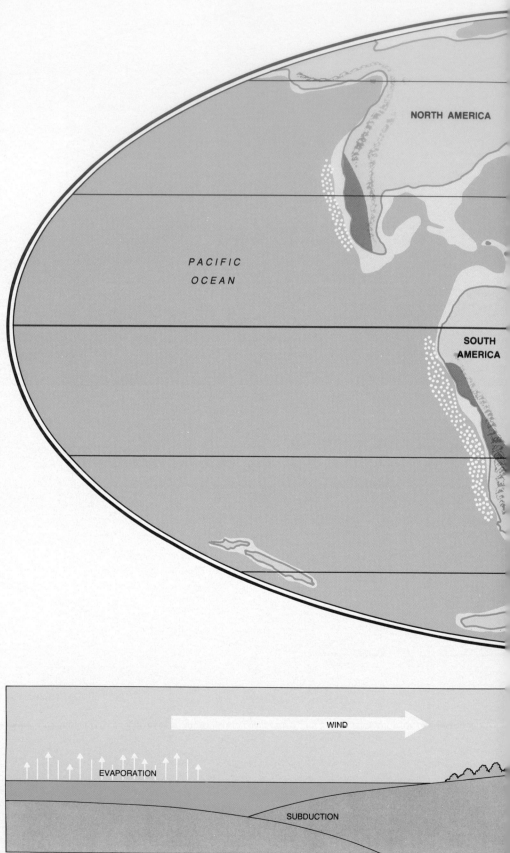

A diagram shows how a coastal mountain range, created by the subduction of one tectonic plate beneath another, produces a rain shadow. As moist air rises to clear the range and encounters lower temperatures aloft, its water vapor falls as precipitation, saturating the windward slope. Because the dry air warms as it descends the lee side, it absorbs moisture more readily and further desiccates the land in the rain shadow.

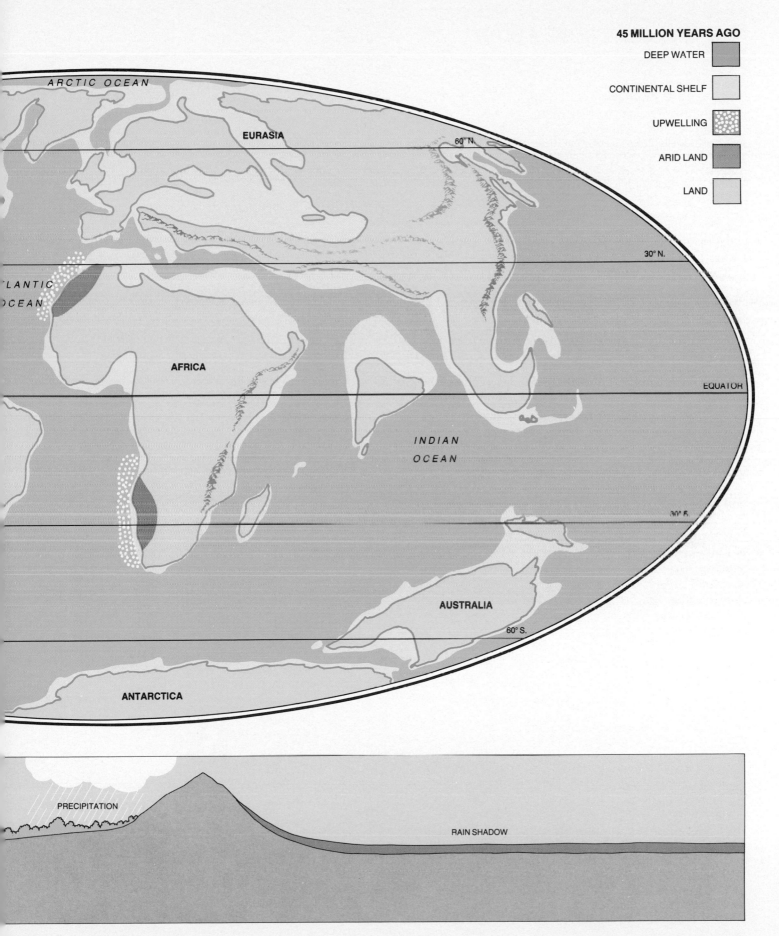

45 MILLION YEARS AGO

DEEP WATER

CONTINENTAL SHELF

UPWELLING

ARID LAND

LAND

ARCTIC OCEAN

EURASIA

60° N.

30° N.

ATLANTIC
OCEAN

AFRICA

EQUATOR

INDIAN
OCEAN

30° S.

AUSTRALIA

60° S.

ANTARCTICA

PRECIPITATION

RAIN SHADOW

The Role of Continental Scale

India's 80-million-year journey northeastward from Antarctica — at the tectonic speed of 12 inches a year — ended by 15 million years ago, after the subcontinent collided with Asia. With the addition of this large land mass, central Asia lay thousands of miles from water, beyond the reach of even the mighty summer monsoons that originate in the Indian Ocean and deposit hundreds of inches of rain inland every year. Moreover, the monumental collision of tectonic plates thrust up the massive Himalayas along their border, casting a rain shadow that also played a role in denying rainfall to the interior. The result of these geologic movements was the resurgence in the heart of Asia of a mammoth arid region, the precursor of today's Gobi desert.

At about the same time, Australia moved northward into the subtropic high-pressure area. The size of the island continent, along with the appearance of upwelling currents off its west coast, made desertification there inevitable.

In South America, on the other hand, the extent of arid land decreased dramatically with the disappearance of the Patagonian desert. While the reasons for this development are unclear, it is thought that an unusually warm and moist climate during the period, combined with a slight lowering of the Andes by weathering, allowed enough saturated air to cross the previously impassable mountains to moderate the interior climate.

A diagram in cross section of a slice of Asia extending 5,000 miles from the Arabian Sea to the Pacific Ocean (indicated in red on the map above) illustrates central Asia's isolation from water. The 1,500-mile breadth of the Himalayan chain further shields the arid lands of Mongolia from moisture-saturated westerlies.

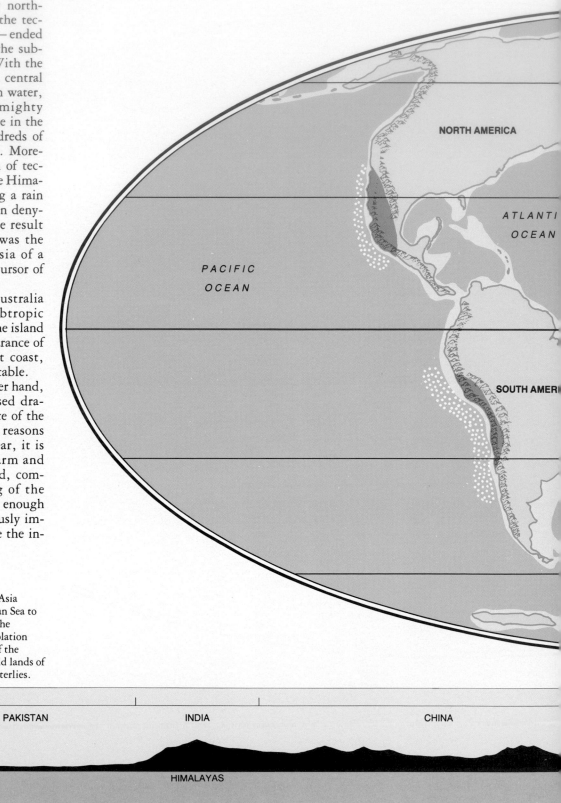

NORTH AMERICA

ATLANTIC OCEAN

PACIFIC OCEAN

SOUTH AMERICA

| Arabian Sea | PAKISTAN | INDIA | CHINA |

HIMALAYAS

0   1,000   2,000

SCALE IN MILES

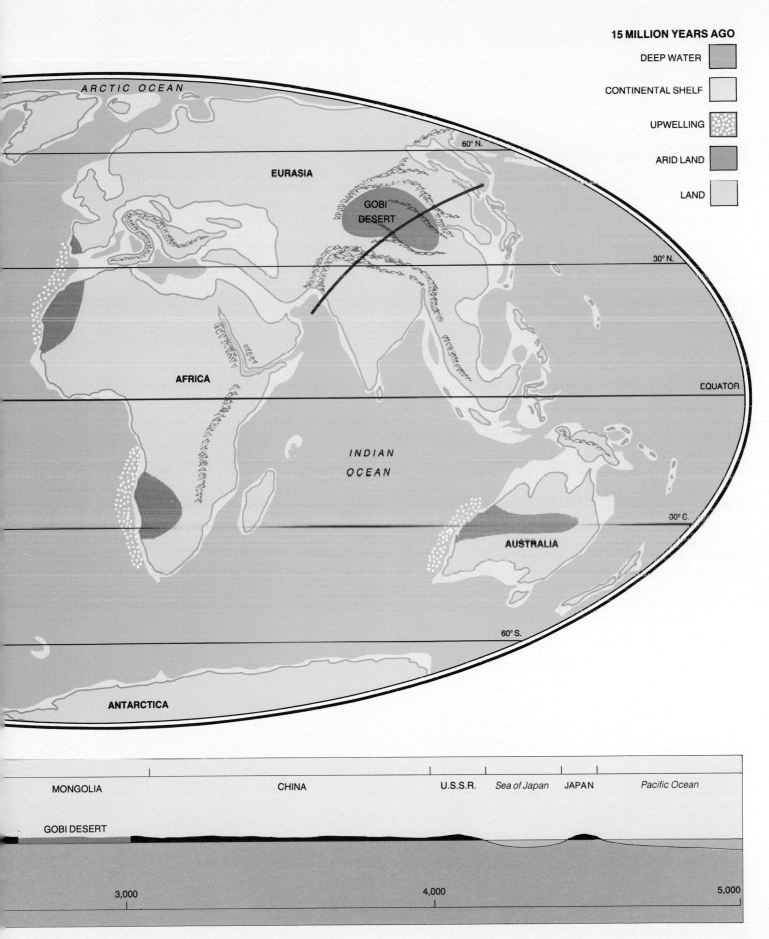

15 MILLION YEARS AGO

DEEP WATER

CONTINENTAL SHELF

UPWELLING

ARID LAND

LAND

ARCTIC OCEAN

EURASIA

GOBI DESERT

AFRICA

INDIAN OCEAN

AUSTRALIA

ANTARCTICA

60° N.

30° N.

EQUATOR

00° C.

60° S.

MONGOLIA   CHINA   U.S.S.R.   *Sea of Japan*   JAPAN   *Pacific Ocean*

GOBI DESERT

3,000   4,000   5,000

Megadeserts of the Future

When 100 million years of anticipated tectonic movements are projected, along with the consequent redistribution of arid lands, the result is a vision of enormous future deserts sprawling across the Northern Hemisphere. The northward drift of Africa will slam it into Europe and Asia, folding new ranges of mountains along the collision zone and eliminating the Mediterranean Sea to form a single prodigious land mass. The size of the new continent, combined with the rain-shadow effect of the emergent mountains and upwelling off the western coast, will create a huge transcontinental desert. All the major deserts of present-day Africa, Arabia and Asia will merge into a swath measuring perhaps 6,000 miles long by 1,000 miles wide.

Crustal shift will pull Greenland and North America away from the Arctic Sea, permitting warm water to reach and eventually melt the polar ice cap. The meltwater will raise sea level 600 feet, inundating the Mississippi and Amazon basins and the African Rift Valley.

As South America moves northward, its coastal desert, fixed in the dry subtropical latitudes, will appear to migrate southward at the same pace. Similarly, while Australia is carried southward, its desert will seem to move to the north.

One exception to the enlargement of northern deserts will occur on the west coast of North America. As that continent pivots counterclockwise, plate movement will shave off some of the coastal mountains. As a result, present rain-shadow deserts east of the mountains will get more rain and will eventually turn green.

A comparison of present continental configurations with those predicted for 100 million years in the future shows a markedly different arrangement of land under the subtropical belts of high pressure (*red bands*). With far more land in the vicinity of lat. 30° N., the Northern Hemisphere will have a colder, drier climate and more extensive deserts. Conversely, less land and more water in the Southern Hemisphere will contribute to a warmer, wetter climate and fewer deserts.

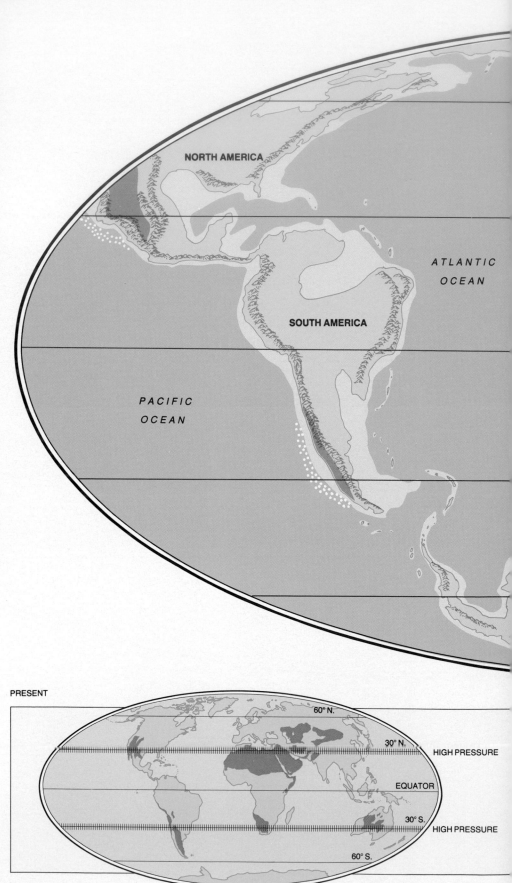

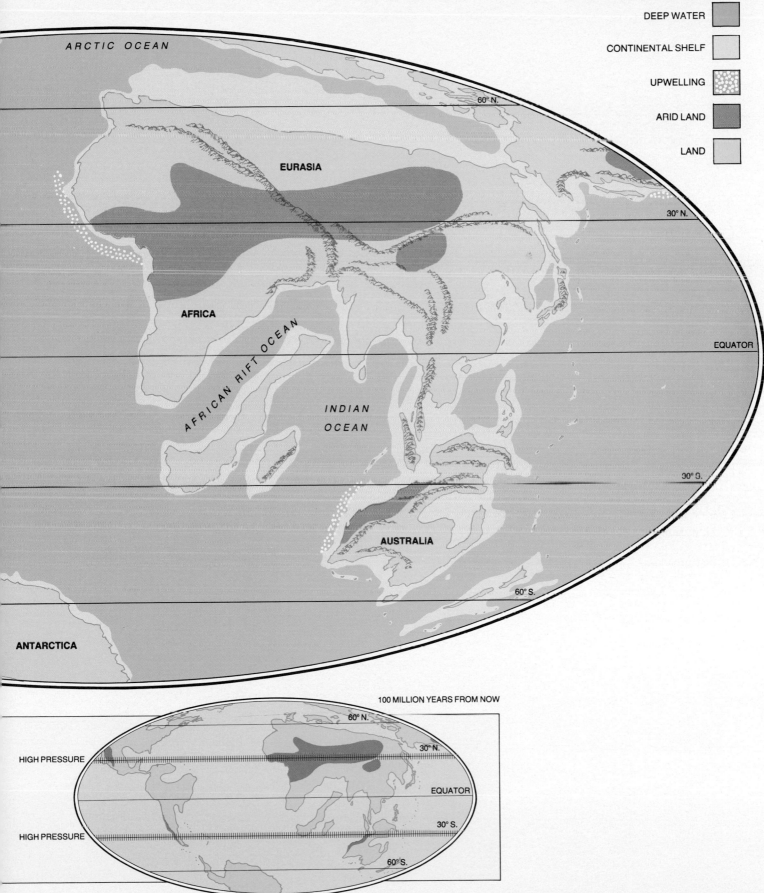

DEEP WATER

CONTINENTAL SHELF

UPWELLING

ARID LAND

LAND

ARCTIC OCEAN

60° N.

EURASIA

30° N.

AFRICA

AFRICAN RIFT OCEAN

EQUATOR

INDIAN OCEAN

30° S.

AUSTRALIA

60° S.

ANTARCTICA

100 MILLION YEARS FROM NOW

60° N.

30° N.

HIGH PRESSURE

EQUATOR

30° S.

HIGH PRESSURE

60° S.

Chapter 3

SURVIVORS IN A SERE WORLD

The famed comedian W. C. Fields, one of the biggest Hollywood stars of the 1940s, assiduously cultivated the image of a dissolute old curmudgeon. He maintained a distinctly sour view of the world and loudly proclaimed his love of alcohol, along with an equal aversion to animals and small children. Yet according to one story, he was caught at least once exhibiting a flicker of sentiment, elicited, oddly enough, by an experience in the arid American Southwest.

It was springtime, and Fields was out strolling near where he was staying when he chanced upon a desert flower in spectacular bloom. He was entranced and felt a tremendous urge to share this beauty with someone. He called his agent in Los Angeles to insist that the man drop everything and drive several hundred miles to see, as Fields put it, "something truly remarkable." The agent arrived next day and, despite his weariness, was immediately dragged by Fields out into the desert to see the wondrous bloom. But on arriving at the site, Fields could find no flower, only an unadorned and rather homely plant. Outraged at the blossom's duplicity, Fields seized his walking stick and thrashed the plant, roaring, "Bloom, damn you, bloom!"

Fields, a master of timing in his own right, had been victimized by timing of another sort, ephemerality, a strategy that many desert organisms have evolved for survival. Ephemeral plants bloom for a day, or somewhat longer, in response to the erratic blessing of rainfall. The strategy is visible in all its glory when, after a brief shower, a barren desert landscape suddenly blossoms in an array of colors: white primroses, blue lupines and yellow desert sunflowers. The business of pollinating the short-lived blossoms is carried on by the wind or by alert and opportunistic insects, then the plants abruptly wither as desiccating conditions return. So effective is this strategy that the largest proportion of annuals in most of the world's deserts is ephemeral.

Given the formidable obstacles to survival in the desert, the mere existence of life seems miraculous. The scarcity of water combined with the tyranny of heat, and in some places, protracted periods of subfreezing cold, make arid lands as hostile as any environment on the planet. In many such regions, including portions of the Namib Desert in southwestern Africa and of the Atacama-Sechura in Peru, life can get no grip at all. Yet where conditions are merely marginal, a surprisingly varied plant and animal community endures. In many cases, extraordinary physical or strategic adaptations, the products of many millennia of evolution, provide a crucial

A climactic event in the 15-to-50-year life of a century plant — the opening of the only flowers it will ever produce — represents merely a midnight snack to this long-nosed bat. In order to survive, desert animals must be alert to such rare instances of abundance among plants struggling to exist on scarce resources.

edge in the battle to survive. The transparent eyelids of certain lizards, for example, permit the creatures some vision even in sandstorms; the mesquite bush has developed the capability of sending down a taproot as far as 175 feet in search of lingering traces of moisture. Ephemerality is one such adaptation, in which the biological clocks of ephemeral plants and animals have been adjusted over time to provide for extended periods of dormancy during drought and instantaneous resurrections to reap the bounty of a sudden rainfall.

Nature photographer Robert Gilbreath chanced upon a plant that exhibited not only ephemerality, but another key strategy of many desert plants. In 1971, during his patrols of the Nevada desert, Gilbreath found a three-inch fungus growing out of rock and sand. Since fungi are normally associated with damp places, Gilbreath decided to photograph the oddity, and in the process he accidentally dropped his lens hood. "Leaning down to pick it up," he wrote later, "I noticed that the ground was covered with tiny plants. On the plants were white specks." He photographed them. When he enlarged the picture, to Gilbreath's amazement, "it showed an exquisite white flower about 1/16 of an inch in diameter with six petals striped with pink, complete with pink-tipped stamens and a yellow pistil."

Gilbreath had happened on the desert's tiniest flowers, thousands of which can occupy a few square feet, unseen by virtually everyone who passes

Highly magnified in the top photograph and shown actual size above, a desert miniflower is attended by an equally tiny insect. Able to survive in the desert because they use little water or energy, the plants emerge from dormant seeds only when conditions are perfect — perhaps once in several years. The blossoms appear and wither away in a few hours.

by. During the next few years Gilbreath spent many hours stretched out on the desert floor, seeking out these miniflowers with a magnifying glass, often seeing the still-tinier insects that pollinate them. He once found a place where four different types of flowers bloomed in a profusion of nearly 200 blossoms per square foot. Returning the next morning to photograph them, he found only two types — the others, ephemerals, having vanished overnight.

Their very tininess provided Gilbreath's flowers with a competitive edge in their harsh world. The smaller the leaves and stems, the less water is needed for sustenance and is lost through transpiration, and the less energy is required to germinate tiny seeds before the limited moisture evaporates. Seed production and germination — the perpetuation of the species — is the ultimate function of all plants. Ephemerals further enhance their chances of continuity with yet another strategy, the manufacture of an abundance of seeds. A stand of one annual, the woolly plantain, has been estimated to produce more than 358,800 seeds per square yard, or almost two billion per acre. Such bounty is hardly superfluous. The seeds may have to wait a year or longer for the next rain, and meanwhile they provide food for scores of insects. Ninety-one per cent of the food stored by harvester ants, for example, is seeds. Nor is it simply a matter of waiting for the first rainfall that occurs; if the seeds germinated and sprouted in response to a mere sprinkle of rain, the plants would soon be left to wither in the sun when the moisture evaporated. This seldom happens; there is evidence that the seeds are coated with a chemical that inhibits sprouting until a significant rainfall washes it away.

The texture and composition of the soil plays at least as important a part as rain — or the lack of it — in determining whether an individual desert plant will survive. Unlike the constantly renewed earth of temperate zones, arid soils lack the enormous enrichment provided by decaying organic matter. But where rain water can penetrate deeply, the desert is often rich in mineral nutrients that are carried back to the surface by water percolating upward from the mineral-laden depths as surface moisture evaporates. It is this question of permeability that is the main factor influencing desert vegetation.

Soil type — and thus permeability — is determined by such things as the nature of the subsurface rock, the amount of weathering and erosion that has taken place, and the topography of the surface. And all these factors are linked to altitude. In the heights of a dry mountain, where the temperature extremes between night and day are severe, the exposed rocks gradually crack and crumble. When the infrequent storms unleash their moisture, it courses down through stream beds in torrents, further pulverizing the rock fragments and sorting them, depositing rocky soils on steeper slopes and carrying gravelly soil down to form alluvial fans below. After many eons of accumulation, redistribution and compression, the finer materials of the alluvial fans form the clayey soils that characterize the lowest reaches of most desert basins.

Such soils become iron-hard when baked by the sun, and most of the rain that falls on them simply drains away. When this runoff water collects to form a shallow lake or pond, excessive percolation triggered by rapid evaporation may bring too much magnesium, sodium or chlorine

As if to compensate for its rarity, a violent rain shower sends torrents of water lashing down on a stand of cacti and shrubs in Arizona's Sonoran Desert.

The storm will trigger a series of desert events: sudden, lush plant growth, cooler weather and a flurry of animal activity.

Bowed by the bounty of a wet season, a desert poplar tree in central Australia shows the strain of carrying on its species' fight for existence. When moisture permits, the tree produces huge quantities of fruit to better the odds of its seeds' survival.

salts to the surface, creating extremely salty soil. But in the coarse-grained soils found in alluvial fans, rainfall soaks in to depths of 16 to 20 inches. There, when protected from the sun and the dry wind by a pavement of pebbles mixed with sand, it remains as free water or forms a fine film that clings to soil particles. Deeply rooted plants can extract both kinds of water. But the finer the soil particles, the more tenaciously water clings to them. Probably the most nurturing soils in arid regions are those that are relatively coarse near the surface but are underlain by finer particles, which tend to keep the water from descending beyond the reach of shallow-rooted plants.

Each of the soil types found from the high ridges to the low basins supports its own uniquely adapted plant and animal community. Joseph Wood Krutch, a New Englander who went out to the American Southwest to chronicle the ways of the desert, noted that specialization when he wrote, "Many a plant, bird and animal sticks almost as closely to his altitude belt as though he were on an island surrounded by water instead of merely in an environment determined by his height above sea level."

This stratification can be readily detected in the plant life one encounters descending a mountain slope in the American Southwest. Similar discrete bands can be found in other deserts, and in fact it sometimes happens that comparable strata produce plants native to widely separated deserts that are all but indistinguishable from one another. Only recently did taxonomists discover that nearly identical creosote bushes native to North America and to Argentina actually belong to separate species.

Barren for most of the year (*above*), the ocotillo responds to rain with a sudden bristling of tiny leaves (*right*). Found in America's southwestern deserts, the shrub will lose its leaves just as quickly at the onset of drought, replacing them after every summer rain.

The highest rocky hillsides of the Southwestern American deserts, where the soil is shallow and drains very quickly, are the domain of the ocotillo, or coachwhip. The plant consists of a cluster of thorny unbranched stems up to 15 feet in height, which sway in the wind and clearly resemble their namesake. But let there be a substantial rain at any time of year and the ocotillo's stems are quickly covered with hundreds of tiny leaves, which, as the moisture evaporates from the soil, turn brown and drop off within a couple of weeks. Spring and autumn come often for the ocotillo, but in April or May of each year it experiences one true spring. Then, even if there has not been enough rain to produce leaves on the coachwhip stems, buds emerge at the tips and burst into clusters of brilliant red flowers. The desert slopes, it has been remarked, appear to be lit with the flames of hundreds of candles. Krutch felt that this curious plant, the only one of its genus in the United States, "expressed the spirit of the Sonoran desert — one which combines oddness of form and habit with the courage to flourish under seemingly impossible conditions, and which combines also the defensive fierceness of thorns with the spectacular, unexpected beauty of brilliant flowers."

Dotting the same coarse, shallow soils that nurture the ocotillo, there appear curious rosettes of thick, pointed leaves. If it has been dry for a long time, the leaves appear to be nearly dead and point directly upward, clustered closely together to avoid as much of the sun's searing heat as possible, while in more kindly times, they spread out. This is an agave, called the century plant because after many years — between 15 and 50, not in reality

97

a full century — a single stalk emerges from the roseate leaves, grows perhaps 15 feet tall and produces spectacular clusters of light-colored flowers; the stalk then dies back, as does the entire superstructure. The agave has adapted to the particularly dry upper altitudes with a slowing of its life processes that permits it to be patient about rain.

The elevations below the hillsides, along the upper reaches of the alluvial fan, are the domain of the cacti — the barrel cactus, the prickly pear, the cholla and the giant saguaro, which reaches heights of 50 feet and lives as long as 200 years. No plant universally epitomizes the desert like the cactus, yet the cactus is found almost exclusively in the New World. Other succulents, including many species of euphorbia that closely resemble the cactus, are found in the arid lands of the Old World.

One reason for the success of the cactus is its efficient strategy for finding and storing water. Cacti have evolved shallow roots that spread out laterally over a wide area and absorb any rainfall before the water either sinks out of reach, runs off or evaporates. Since rain is infrequent, the water thus obtained must be conserved. Rain water is transported quickly through the roots and the vascular tissue of the stem into the fleshy green interior of the cactus, where photosynthesis takes place and excess moisture is stored. The enormous saguaro cactus stores from six to eight tons of water in its trunk. The moisture-retaining ability of the cactus family has been credited with keeping alive any number of people stranded in the desert. In one instance, a Marine Corps pilot forced to parachute from his burning plane into the Arizona desert later wrote, "I wouldn't be here today if it weren't for the barrel cactus." Lieutenant Edwin Zolnier wandered in the desert for five days before being rescued. His desperate search for water was unavailing,

Swollen with stored water, a saguaro *(above)* in Arizona's Sonoran Desert seems about to burst from the strain of its precious hoard. After a heavy rain, a large saguaro can soak up a ton of water, doubling its girth. A woody skeleton supports the cactus, which can reach a height of 50 feet, while the pulpy tissue surrounding the core *(below)* holds water like a sponge.

but he found a barrel cactus "and chewed the water out of large chunks cut from the living plant." Repeatedly during his ordeal, Zolnier hacked open a cactus, sucked the juice and rubbed his body with the moist pulp to stave off dehydration.

In addition to gathering and storing the scant moisture of their environment, cacti exhibit several strategies for warding off the effects of constant, burning heat. A waxy cuticle, or skin, discourages evaporation. The fluted ridges, bumps and nipples on the surface of cacti represent a kind of compromise adaptation that allows the maximum amount of sunlight, which is necessary for photosynthesis, to hit the stem while at the same time providing small but significant recesses to enhance cooling. In the heat of day, the cactus closes its pores so water cannot escape, using and recycling within its system the carbon dioxide it utilizes for photosynthesis. Then, to replenish its supply, it opens its pores to the cool night air.

Most cacti bristle with spines, which offer important advantages over the leaves of the typical temperate-latitude plant: Spines permit heat to be given off, require little energy to maintain and expose a minimal amount of surface to desiccating winds. The spines of some species are lighter in color than the stems and serve to reflect some of the sun's heat. Spines also discourage animals from eating the plant, and when they point downward, they concentrate the runoff of a sprinkling rain into droplets that fall near the base of the plant, where its thirsty roots can gather the moisture. For additional protection, the surfaces of many cacti as well as other desert plants have numerous white hairs that reflect heat away from vulnerable growing tissue.

Like the cacti, other midslope desert plants have evolved efficient survival strategies. An example is the succulent aloe plant found in Africa. The aloe stores enormous quantities of water in its stem, which swells after a rainfall — one variety achieving a height of 60 feet and a diameter of up to nine feet. Another inhabitant of desert slopes, specifically those of the sere lands of Baja California, is the boojum tree, which resembles a huge, single, wizened carrot thrust up from the ground. To conserve energy, the plant produces a light covering of small leaves on the thin branches only after a rain.

Downslope from the realm of the large cacti and boojum trees, the soil of the lower portions of the alluvial fan, having been carried by wind and water, becomes finer, and consequently the vegetation changes. Here, shrubs such as the odoriferous creosote bush dominate in the American Southwest. A number of small adaptations have equipped the creosote for desert survival, including roots that excel in extracting moisture from soil and leaves whose development can be arrested during drought, then renewed when conditions improve. The creosote, with its extensive root system, crowds out less opportunistic competitors, such as grass, and is considered a pest by ranchers whose cattle must range far and wide for sparse tufts of forage.

Lower on the slope, one finds woody shrubs like the mesquite, whose roots can reach down 175 feet in search of water. Finally, in the closed basins where the finest silt is deposited, runoff collects and the soil is salty, the desert floor will be covered with pickleweed and saltbushes. These bushes are among the few shrubs that can live in highly saline soils because their own tissues become as salty as the soil in which they grow. The eventual decomposition of their leaves and stems tends to make the soil near the plants even more saline.

Along the arroyos that cut through these bottom lands, desert willows and cottonwoods thrive, their root systems delving as far as 40 feet down or pushing horizontally through the dry soil to the wetter subsoil beneath the arroyos. Nevertheless, trees are a rarity in deserts, where growth is more profitably expended in the search for water underground, and vegetation rarely exceeds 30 feet in height. Even in less arid places trees, and for that matter even shrubs, must be widely spaced to assure each plant an adequate share of the available moisture. Some plants manufacture toxins that seep into the soil, preventing the sprouting of competitors and thus protecting their territory—a kind of chemical security system.

In fact, much of the competition among plants for scarce desert moisture takes place underground. Most plants have developed enormous root systems at the expense of aboveground foliage and must sustain those roots with less chlorophyll-bearing surface to carry on photosynthesis—the job done for most other plants by leaves. There is evidence that despite their size, the roots of desert plants utilize respiration rates that are much lower than temperate-region plants and thus require less energy to maintain. Desert plants must also resist overheating, so they tend to have small, waxy, often resinous leaves in which the cells are more compact and the pores tighter. The leaves are less green than their temperate-latitude counterparts and, like the leaves of Australia's eucalyptus trees, shade toward gray, a

Protruding from its rocky terrain, a velvet rosette demonstrates one of the many ways desert plants adapt to the constant shortage of water; the downy filaments covering the leaf clusters for which the plant is named help to extract moisture from the air.

color that reflects much of the sun's excessive energy. A similar color adaptation can be seen in the whitened, sharp-spined clumps of grass that dot Australia's northern grasslands.

Biologists point out that there is no plant — particularly no desert plant — that is optimally adapted to its particular set of conditions. If one were, it would probably be a likely candidate for extinction because conditions change rapidly in arid lands. Long droughts, shifts in the timing and amount of rainfall, changes in the nature of soils and the direction of watercourses — all these factors are at work in the mosaic of ecosystems that constitute arid lands. In order to survive, desert plants must make continual adaptive compromises, taking a little less of this in order to get enough of that: spines instead of leaves, huge root systems at the expense of aboveground foliage and stems, long spells without leaves, shortened periods of photosynthesis. Given the difficulty of sustaining life, it costs the desert plant relatively more available energy to produce its compact leaves, with their special ways of accomplishing their tasks, than it costs a maple to produce its abundant, expansive leaves. But the compromises must be made if the plant is to survive the metronomic reoccurrence of the sun's fierce rays, the erratic availability of water and the punishing fluctuations in temperature.

Great temperature variations are frequent, even daily, in many desert regions; a midafternoon reading of 120° F. may be followed, just before

Shortly after a rainfall, a desert flat in Western Australia is suddenly bedecked with brilliantly colored blossoms that are unsupported by stems or leaves. The plant has made use of the rare moisture to produce the parts most essential to preserving the species — the flowers. If sufficient rain follows, the rest of the plant will appear.

Another denizen of the Namib Desert, this *Cissus macropus* belongs to the grape family. The plant stores water in its thick trunk, which is insulated with pale bark that reflects heat.

Found only in Africa's Namib Desert, these long-lived welwitschias obtain the moisture they need from fog that rolls as far as 50 miles inland from the sea. During their thousand-year lives, the plants' two leaves, shredded by the wind, create about 25 square yards of surface area capable of absorbing water vapor.

sunrise, by one of only 60° F. A plant that is well adapted to great heat and dryness may not be able to live in certain hot, dry areas because it cannot stand the cold that often follows the setting of the sun. With little cloud cover to help retain heat, the desert may radiate back into the atmosphere as much as 90 per cent of the heat energy received that day from the sun. Moreover, with the sunset, air at the colder altitudes of nearby highlands may become heavier than the air below it and roll down onto the desert floor like an invisible fog, dropping temperatures as much as 50° F. The extremes of the sun's heat are moderated somewhat on the north-facing slopes of a mountainous desert in the Northern Hemisphere (the southern slopes in the Southern Hemisphere). The difference is not necessarily much, but enough that the south slope may have quite a different assemblage of plants than the north-facing slope, even though the altitude is the same.

There may be no perfectly adapted desert plant, but there should be some sort of prize for *Welwitschia mirabilis*, certainly one of the most unprepossessing — and successful — examples of vegetation one could imagine. While almost all arid-region plants lack the luxuriant beauty of normal herbs, shrubs and trees, many exhibit their own particular loveliness and grace — that of the spectacular varieties of cactus flowers, for example, or the supplicant arms of the Mojave Desert's giant yucca, which the Mormons called the Joshua tree. But welwitschia looks for all the world like a pile of dirty laundry. It lives in the Namib Desert of South-West Africa, a place where moisture comes mainly from the fog that rolls inland from the sea, often as far as 50 miles. Rain is erratic and infrequent — averaging about two inches a year.

Seeds produced by the welwitschia germinate within three weeks of being wet, but this is the only thing the plant does quickly. After sending down a root that resembles a carrot, welwitschia puts forth just two small, leathery leaves. As the taproot grows, penetrating as far as 60 feet down in quest of moisture, the plant's two leaves grow at the leisurely rate of four to eight inches a year, splitting and shredding in the buffeting wind until there exists a dense five-foot-high tangle of ragged foliage. It takes the plant 25 years to put forth its first flower. One of the few plants that survives in the Namib, some welwitschias have been estimated through radiocarbon dating to be 2,000 years old.

Yet even such plants as these are virtual youngsters when compared with the premier survivor of them all, a bizarre inhabitant of the cold desert, where all life faces the double jeopardy of deep cold as well as drought. By all appearances, the bristlecone pine can hardly be said to thrive, yet it endures scant oxygen, bitter cold and ceaseless dry wind at altitudes of 8,000 to 10,000 feet in the mountains of the American West, where it finds anchor in barren rubble patches of rock. Despite such adversity, the seniors among bristlecones are the oldest living things on the face of the earth. The trunk of one that was felled displayed 4,900 annual rings; a test-boring of a living bristlecone in the White Mountains of California showed it to be 4,600 years old. These ancients exhibit anything but a robust old age. Gnarled, unshapely dwarfs, they appear to be surviving relics of another time. Much of their wood — including, in many cases, the trunks — is dead, though stubby live branches may still protrude from their bases. Moreover, these trees are unable to survive anywhere but in their austere

Looking more dead than alive in their brutal high-desert environment, a stand of bristlecone pines forms an eerie landscape in California's White

Mountains. Despite the hardships that punish these trees, some have endured for more than 4,000 years; they are the world's oldest living things.

high-desert habitat, having lost the potency to develop as a stronger, more flexible strain that might compete with other trees and thus extend their range. They appear, in all, to be at a dead end on the evolutionary road.

The harsh evolutionary trials that eliminated most plant species from the hostile environment of the desert have had the same effect on animals. The parallels are most striking in the various kinds of aquatic creatures that, after long periods of deathlike inactivity, burst into abundant life in the temporary briny lakes that are formed on playas by a substantial rain. The tiny eggs of various crustaceans, such as brine shrimp, fairy shrimp and tadpole shrimp, may have been dormant for as many as 50 years in the salty crust. With the coming of water, they hatch into microscopic larvae, feed on similarly rehydrated algae that bloom with the newfound moisture and rapidly mature. The adults mate and lay their eggs, completing their life cycle before the salty water evaporates. And wherever around the world this phenomenon takes place, even if it occurs only once or twice a century, waterfowl somehow find the temporary lakes and gorge on the teeming life.

The strategy of the spadefoot toad of the American Southwest may provide the basis for the assertion, in the folklore of the Navajo Indians, that frogs and toads fall from the sky with the rain. With the onset of drought, adult spadefoot toads dig their way backward into the sand and remain dormant for as long as eight or nine months. When they detect a summer downpour, the toads emerge and mate, laying their eggs in the puddles. Within two weeks, tiny tadpoles emerge as if begotten by the puddle (page 110).

A number of other species also make use of opportunistic breeding patterns that provide the greatest chance of reproductive success. To produce young when food is scarce may jeopardize an entire generation and deplete the strength of the mother. Animals such as the Asian gerbil reduce or stop breeding entirely when the rains fail to occur. Similarly, the reproductive cycle of some birds, which in other environments is regulated by the lengthening days of spring, responds in the desert to the stimulus of rainfall, with its promise of plentiful vegetation, readying the bird quickly for productive breeding.

For desert animals, survival requires more than simply enduring the long wait for rainfall. To thrive in their austere world, they must find food, avoid predators and deal with competition from any other creatures in their ecological niche. The adaptations tend to be as extreme as the conditions that prompt the changes, especially where water conservation is concerned. And no better example of such an adaptation can be found than that of the camel.

It has been said that the camel looks like a horse that was assembled by a committee. If so, the committee knew a great deal about the problems of living in arid lands, for the camel is the quintessential desert animal. Of course, camels were not in any sense designed but came to be what they are today by the adaptation to changes in their environment over long periods of time. They first appeared during the warm, humid climate of the Eocene Epoch in what is now North America about 40 million years ago. They were relatively small animals — perhaps pig-sized — with short legs and four toes. They descended from a small ungulate, or hoofed animal, that is

1,000 FEET   80° F.

7 FEET   100° F.

3.5 FEET   122° F.

SURFACE   167° F.

5 FEET   86° F.

7 FEET   80° F.

Although the temperature of a desert's surface may exceed 150° F. by midday, relief is just moments away for most animals. A few feet underground, ants and jerboas enjoy surroundings that may be nearly 90° cooler — as does the vulture soaring 1,000 feet aboveground. Just by climbing a small bush, a monitor lizard finds air that is 45° cooler. And the camel's long legs hold its body about four feet above the surface, where temperatures are almost 60° lower.

also an ancestor of the boar, hippopotamus, deer, giraffe, antelope and bison. Numerous camel types evolved, but by about 25 million years ago, the typical example resembled today's llama; it had two toes instead of four and its legs were longer (adaptations that would have made the animal faster and more agile) and the neck was extended to facilitate grazing. In one variety, called Alticamelus and often referred to as the giraffe camel by paleontologists, the head was a full 10 feet above the ground, permitting the creature to feed among the leaves of tall trees.

Camels were restricted to the North American continent until about 2.5 million years ago. Then, with the onset of the ice ages, they moved into South America and across a new intercontinental land bridge that spanned the Bering Strait, into arid regions of Asia and Africa. Probably at about that time, the camel developed flattened, spreading pads on its feet, which eased walking through sand, and also acquired its characteristic hump of stored fat.

Inexplicably, the camel eventually disappeared from its native continent, but it flourished in Asia and eventually became the universal symbol of life in the desert.

Legend dating back 1,900 years to the Roman naturalist Pliny the Elder held that camels stored water in their stomachs; that was why, it was said, they could go for so long without water, and why they often drank a great deal of it before a caravan began. So firmly entrenched was this belief that travelers who became stranded in the desert were known to slit open the stomachs of their camels and drink the vile-smelling liquid inside. Modern analysis has shown the stuff to be, not water, but very salty digestive juices. Still, like any liquid, it would serve to sustain life for a short time.

In fact, a camel can get along without water for as long as 10 days, depending on the amount of activity and the weather, thanks to a wide range of superb adaptations. A camel's woolly hide insulates the animal against the heat and slows evaporation when it sweats, providing maximum cooling for the water lost. Its long legs keep the camel's vital organs well above the scorching desert floor where temperatures may be considerably hotter than the surrounding air. Furthermore, although the camel's legs are adapted for speed, the animal typically moves in an energy-efficient kind of slow motion.

Nevertheless the camel does lose water. And the true secret of its ability to endure an extraordinary amount of dehydration is its body fat. From this fat, which is concentrated in the hump rather than spread through the body where it would hinder the release of body heat, the animal draws not only energy but water. Every ounce of fat oxidized yields slightly more than an ounce of water; thus a typical hump of about 90 pounds represents a source of nearly 45 quarts of water. As the fat is converted to energy and water, the hump shrinks visibly; the camel can lose up to a third of its body weight without harm. One camel is reported to have gone without water for eight days, losing 227 pounds. When given water it gulped down 27 gallons, and within hours the water was evenly distributed throughout its body. More slowly, fat was redeposited in the hump.

The camel species that inhabits the hot deserts of Africa, the Middle East and Asia — the dromedary — has one hump and ranges in color from brown to white. It probably evolved from the Bactrian camel — a shorter, stockier, darker and shaggier two-humped species — whose more massive body and dense winter coat help the animal withstand the frigid climate of the cold deserts of Asia. Camels are so highly valued as riding animals and beasts of burden in these dry lands that they have been almost totally domesticated. There are few wild dromedaries, except for those that stray (they have a very poor sense of direction and easily become lost). About 1,000 wild Bactrian camels live in the Gobi, carefully protected by the Chinese and Mongolian governments.

Inevitably, people exploring or working in other arid places thought about how much easier their tasks would be with the camel's help. In 1875, Ernest Giles used a herd of 24 camels to cross 2,500 miles of Australian Outback from Port Augusta to Perth. The journey took five

The Frenzied Mating of the Spadefoot Toad

For most of the year, spadefoot toads cope with life in the Sonoran Desert of Arizona by avoiding it, remaining dormant in burrows. But as midsummer approaches, the toads respond to some internal cue and inch closer to the surface, occasionally emerging at night to feed. Eventually, a heavy downpour occurs, and the vibrations of thudding raindrops signal the toads to emerge and begin their frantic mating season.

Obeying a timetable imposed by the brief presence of life-sustaining water, the females lay their eggs — as many as 1,000 each — within hours of the storm, in small pools of rain water. From then on, survival of the young becomes a race against the elements as the puddles evaporate, growing ever smaller under the hot sun. If a pool vanishes too soon, the eggs and tadpoles will die with it.

To survive, the young must mature with remarkable speed; in as few as nine days after fertilization of the egg, a spadefoot may already have completed the tadpole stage to become a fully formed toad. Like their parents, the young toads then quickly burrow underground, where they will lie dormant, awaiting next summer's rainfall.

A female spadefoot toad emerges from her burrow to mate. But first she devotes herself to gorging on insects; in her brief stay aboveground, she must eat enough to sustain her through at least nine months of hibernation.

A male spadefoot toad fills the night with his calls as he lures a mate to his muddy pool. When he locates a female of the same species, mating and egg laying will continue all night.

After mating, an adult spadefoot toad retreats into the cool, wet mud to escape the sun. It burrows in backwards by using horny "spades" on its hind feet to shove the mud aside.

About 36 hours after the eggs are laid, hordes of tiny tadpoles emerge. Their first few days are threatening ones as they face challenges from predators and the fierce desert sun.

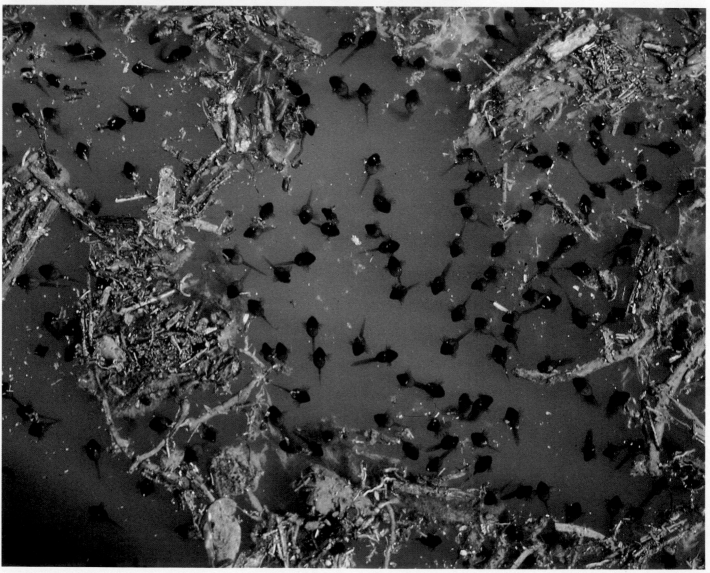

Only days old, these tadpoles have already sprouted legs. Their thrashing has thickened the increasingly shallow water into mud.

Encased in mud, doomed tadpoles whose puddle has evaporated are easy prey for a band of ants. Danger lurks beneath the mud as well; large, grublike horsefly larvae have been known to latch onto the toads with sharp mouth hooks and drain their body fluids.

One of the surviving few, this youngster has nearly reached toadhood. For the next few days, not yet able to burrow as adults do, the toad will search for shade and food as it prepares to join the adults in their long hibernation.

months and passed through one waterless stretch of 325 miles in an area Giles described as "utterly unknown to man and as utterly forsaken by God." After a two-month rest, Giles returned by an even more difficult northern route to Adelaide. Camels became the accepted mode for opening up Australia's deserts and were used in 1939 by the Madigan expedition, which explored and mapped the last remaining unknown piece of the continent, a region in the Simpson Desert in central Australia. Despite their success as pack animals, no demand emerged for domesticated camels in Australia, but a feral herd of their descendants still roams the Australian desert.

Attempts to reintroduce the camel to its ancestral home, North America, were even less successful. In the 1850s, Major Henry C. Wayne, an assistant to the deputy assistant quartermaster of the United States Army, managed to convince his superiors that camels would be ideal for transporting goods and matériel across the arid lands of the Southwest. Secretary

A lithe, small-limbed lizard called a skink demonstrates a vanishing act it uses to escape predators or the scorching surface of a Moroccan desert. The enlarged scales on its toes and its shovel-shaped head facilitate rapid burrowing through the loose sand.

of War Jefferson Davis, whose interest in the proposal no doubt arose from his years of service on the remote Western frontier, shepherded the idea through Congress, which eventually provided the money for a shipload of 34 dromedaries that arrived in Texas in 1856, and another of 41 animals a year later. The camels trekked without difficulty from Texas to California, content to eat cacti, mesquite and greasewood (forage that no horse or mule would touch). The camels carried heavier loads than the accompanying draft animals and finished the trip in better health. Technically, they were a success.

The problem was one of human relations. Everyone who came into contact with the camels hated them. The animals were stubborn, smelly, noisy (they groan bitterly when being loaded and unloaded) and vicious, often biting their handlers. Even civilians who did not have to deal with the beasts resented their presence; Brownsville, Texas, passed an ordinance banning the malodorous animals from their streets. Mule handlers considered it beneath their professional dignity to deal with camels. Even Army horses balked at being stabled with them. Nonetheless, the camels saw four years of military service before the outbreak of the Civil War, when the animals fell into the hands of the Confederates, who had no better luck with them. After the War, a few private enterprises experimented with camels, but railroad lines snaking across the Southwest soon rendered animal caravans unnecessary. As late as 1920, however, stories were being told of some lonely homesteader being startled out of his wits by a feral monster peering in his window. But for all practical purposes, once again the camel had disappeared from North America.

Camels and many other desert creatures are capable of sustaining themselves without any free water at all if there is sufficient pasturage around. Seeds, stems and fallen leaves are about 50 per cent water, and fresh vegetation, fruits and the tissues of succulents contain far more. From such sources many desert herbivores derive all the water they need, and since their bodies consist of up to 75 per cent water, they in turn supply predators with needed water.

Those creatures that cannot find, or eat, vegetable matter and cannot get along without water, must go to even greater adaptive lengths to obtain and preserve moisture. In southwest Africa's Namib Desert, where rainfall is as rare as in much of the Sahara, there is plenty of moisture — but it is present only in the foggy atmosphere. Certain darkling beetles can take advantage of the fog because they are what used to be called cold-blooded, more properly "ectothermic," which means they gain their body heat from external sources, chiefly the sun. In the morning when the beetles are colder than the surrounding atmosphere, they perch on the tops of dunes, facing the

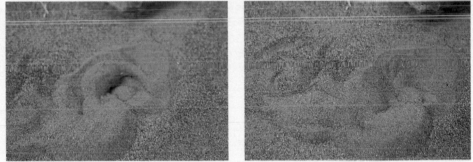

incoming fog, their back legs up on the crest and their forelegs angled down the seaward slope of the dune. The fog condenses on their cool backs and the droplets run down to their mouths. Sand vipers accomplish the same thing by coiling on the fog-shrouded dunes and licking the moisture off their bodies as it forms.

Successful dune dwellers and other inhabitants of sand deserts must traverse a shifting, unstable terrain. To cope, some creatures have evolved unique forms of locomotion. One of the most dramatic is that of the sidewinder rattlesnake. Instead of slithering forward, as do most snakes, it moves by looping its body sideways. Anchoring itself in the sand at two or three points, it arches its body to the side, giving it a slippage-resistant diagonal motion, which has the additional benefit of reducing the snake's contact with the searing sand.

Similarly, on several continents, including Australia and Africa, a family of lizards known as skinks have evolved with reduced limbs or even no limbs at all. This streamlining permits them to live a totally underground existence, swimming through the sand in search of insects or other prey. Other lizards have evolved enlarged scales on their toes, which, like snowshoes, provide traction in loose sand; still others have developed webbing between the toes to achieve the same purpose. The shovel- or wedge-shaped heads of some lizards presumably facilitate burrowing in sand, and transparent eyelids permit the eyes to be closed against blowing sand without impeding eyesight.

A number of desert creatures have benefited from bipedalism — the ability to travel on two legs. The adaptation seems to offer greater speed and mobility for evading predators and appears in the kangaroo rat, in its ecological counterpart in Africa, the jerboa, and, of course, in the kangaroo. Yet one of the most successful desert creatures on two legs is a bird, a form of cuckoo called the roadrunner, whose superbly designed feet have made it a legend in the arid lands of the American West.

Four toes splayed to form an X give each of the roadrunner's feet remarkable gripping power, enabling the bird to run across the hot sands at speeds up to 15 miles per hour. Although the roadrunner can fly, or at least glide, it employs its stubby wings mainly as stabilizers during its sprints across the desert. The sight is impressive — not to say comical — and earned the roadrunner its name in the early days of the American West when travelers were amused to see the bird dashing along a road just ahead of their briskly trotting horses.

Although the roadrunner is no larger than a slimmed-down pheasant, it is more than a match for one of the desert's fiercest predators, the rattlesnake. In a head-to-head encounter, the bird deftly dodges the strikes of the snake, then delivers a powerful, usually fatal, jab to the snake's head with its two-inch beak. Once the snake is dead, the roadrunner gulps it down head first. Because the snake is frequently larger than the roadrunner's stomach, the bird lets the tail dangle from its mouth as it goes about its business, slowly digesting its meal over a period of hours.

Because of its fleetness of foot, the roadrunner can locate food and water over relatively long distances. More sluggish creatures have developed ingenious techniques of food and water storage to provide a margin of protection against want. Harvester ants store annual seeds by the thousands. The honeypot ants of the American Southwest accomplish the same thing in far more exotic fashion. Within their population exists a caste known as repletes. When food is plentiful in moist, cool seasons, other workers stuff liquid food into the repletes' crops, which have internal dividers that separate the auxiliary stomachs from the insects' digestive tracts. When filled until they are swollen and can scarcely move, the repletes crawl to the tops of their underground burrows and hang upside down like round bottles, to be drained of their honey-like contents by the others in the colony during hard times.

An Australian marsupial, the fat-tailed mouse, has evolved what might be called a form of zoological succulence. Like the camel, during periods of plenty the mouse stores fat, not in a hump but in its tail. In times of need the creature oxidizes the stored fat, and its thickened tail becomes thin once again.

For all such exotic strategies against deprivation, the food chain in the desert is for the most part the same as elsewhere: Vegetation supports 10 per cent of its weight in herbivores, which in turn support 10 per cent of their weight in carnivores. What does not fall prey to one creature or another along this chain dies of other causes, becoming detritus, also an important food source, and is processed by scavengers.

But in some parts of the Namib and in South America's Atacama-Sechura Desert, where it is so dry that there is almost no permanent vegetation, the food chain can be significantly altered. During its lifetime, the Peruvian fox of the Atacama-Sechura will feed as a herbivore, a carnivore — its preferred role — and a detritivore, depending on the exigencies of a varying

Its body bloated with liquid nectar, a honeypot ant clings to the roof of its desert colony home in northern Mexico. Worker ants feed these living reservoirs — members of a caste known as repletes — with liquid carbohydrates when food is abundant. During drought, when food sources vanish, the repletes provide the nectar to nourish the colony.

To store heat against the chill of the coming night in the Sonoran Desert, a roadrunner spreads its wings and lifts its feathers, exposing a patch of dark skin on its back that acts as a solar panel to raise its body temperature. By using the sun instead of physical activity to warm itself, the bird saves 550 precious calories an hour.

environment. Unlike an omnivore, which by nature feeds on a variety of foods, the fox is forced by circumstances to alter its habits. When meat is scarce but plants are available, it will eat vegetation; in another season it will subsist on rodents; and at yet another time the animal will feed on detritus, dead sea life that has washed up along the shore. When the fox must subsist exclusively on detritus, it is experiencing a reduction of the normal food chain not found anywhere else except in the ocean's abysses where there is not enough light for marine plants to grow and the denizens must feed exclusively on decaying vegetation that sinks from above, or in the remote depths of caves where troglobites must subsist on whatever comes their way.

The very harshness of life in arid lands brings into sharper focus certain principles of ecology and evolution that apply in other habitats. In the Thar Desert of India and Pakistan, for example, three species of lizard live in the same sandy environment. If they all had the same food preferences, one or two species would ultimately lose out in the competition for limited resources. But one of the lizard types feeds entirely on insects, another feeds on other lizards (including its own species) and the third feeds on plants. Still other lizard species prefer a rocky environment and their feeding habits also differ, one eating insects and small lizards, another large rodents and snakes. This survival strategy, called resource partitioning, is nowhere more clearly to be seen than in the glare of the desert sun.

It is the heat from that ever-present sun that, along with the scarcity of water, presents desert dwellers with their greatest and most constant challenge to survival. Large numbers of desert animals are light in color, often the color of the sand and rock of their environment. This not only serves as protective coloration, allowing the animals to blend into the surroundings and avoid the notice of predators, but tends to reflect heat away from them.

Nevertheless, the danger of overheating is continual. Some animals get enough water in their diets to cool themselves by evaporative heat loss, by panting as do carnivores such as wolves and foxes, or by sweating as do such large herbivores as zebras and gazelles. Other animals have developed radiators, such as the outsized ears of the jackrabbit and many desert foxes, which increase the amount of surface without greatly adding to the animal's volume. The thin bodies and necks and long, slender legs of the dorcas gazelles in Africa and Asia achieve the same end.

Burrows provide daytime refuge for a great many desert inhabitants that forage for food at night, an adaptation known as nocturnalism, which is far more prevalent in the desert than in any other environment. This is the heat-avoiding life style of numerous desert rodents, such as the kangaroo rat and its ecological equivalent, the jerboa. A lizard makes use of variable body temperature to adapt beautifully to daily and seasonal heat fluctuations in the desert. The sun's heat vitalizes its metabolism when activity is necessary, but by crawling into a cool burrow, the lizard slows its metabolic rate during adverse conditions and thus conserves water and energy. Certain species of lizard are often found in the morning lying perpendicular to the sun's rays, but as the day heats up they lie parallel to the rays, presenting the smallest surface to its direct radiation.

Numerous desert rodents and some birds find protection against long dry

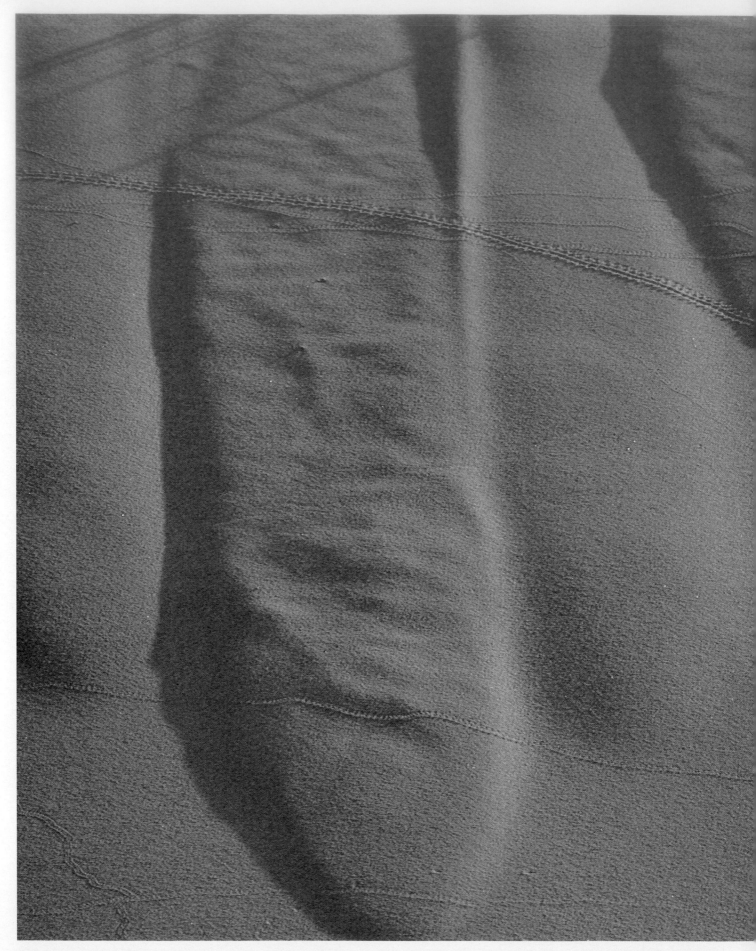

Early-morning light bathing a sand dune reveals traces of recent visitors — beetles and, in the case of the larger trails at right, lizards. Many small desert

creatures are nocturnal, emerging from burrows only when the night hides their activities and cools the sand.

periods in a practice known as torpor; they slow their metabolism until they enter a warm-climate state akin to the hibernation common to cold environments. What moisture there is in the soil lies below the surface, and numerous desert animals reside for months in cool burrows.

Of all the desert creatures with their diverse survival strategies, none seems so admirably adapted as the poor-will, a resident of the cold deserts of North America. A nocturnal bird, the poor-will starts feeding in the cool of the evening during the spring, summer and fall, gorging on insects that whir about in the dark desert air. Manipulating its rounded wings to produce a quick, darting flight pattern much like that of a bat, the poor-will flies with its outsized, shovel-shaped bill and gullet wide open, enjoying its feast without having to alight. By day in the warm months it can seek the comparative cool of its nest.

Thus spared heat and hunger, able to seek out water when needed, the poor-will thrives admirably in the cold desert for well over half the year. But until quite recently no one knew what the poor-wills did during the winter, or even where they were, for in the coldest months there seemed to be no poor-will activity. It was assumed that the birds, like other species in the family, simply migrated to less-hostile climes. Hibernation was not considered to be a possibility, because ornithologists knew that the metabolic requirements of birds were too high to be significantly slowed. Some birds must eat or drink their body weight several times over each day simply to stay alive, and thus it seemed impossible that any bird could throttle back its metabolism far enough to live through hibernation.

The finding of a stiff, cold poor-will in a rocky niche of the Chuckawalla Mountains in the Mojave Desert one late December day seemed to confirm

An Australian frog creates a moist world of its own for protection from the desert's aridity. The frog lies dormant in its burrow most of the year, sheathed in a layer of skin that retains body moisture, and reemerges only during infrequent rains. With its unique cocoon, this frog loses water at $\frac{1}{10}$ the rate of unprotected frogs.

the evidence against hibernation. This particular creature, pulled from its nook by a naturalist who chanced upon it, felt unusually light, as would a bird that had starved to death or become desiccated after perishing accidentally. The scientist replaced it, then returned 10 days later to show it, as a curiosity, to a friend. Clearly the poor-will had failed to fly south ahead of bad weather and had been killed by the cold. But when one of the researchers took the bird in hand to hold it for a photograph, the poor-will spread its wings smartly and flew away — to perch in another niche away from the meddlesome humans.

Subsequent research on the bird, which had returned to roost in the same niche, late the following fall proved that the poor-will did, indeed, hibernate — with astounding virtuosity. Taken once again from its cubbyhole in the cliff, the poor-will registered no audible heartbeat when checked with a stethoscope and left no breath mist on a mirror held to its nostril. Its body temperature had dropped from a normal 108° F. to a deathlike 64° F. Taken to a laboratory where it was kept refrigerated, the poor-will showed no signs of life for three months — whereupon it awoke, healthy and lively as ever.

Even more fascinating was the fact that when the awakened bird was popped back into a refrigerator, it suffered no harm and showed no distress. It simply dropped back immediately into its winter sleep. Mammals cannot do this. Normally they require weeks to slow their metabolisms for hibernation. Yet among the high cliffs of a cold desert, the warmth of Indian summer may vanish within hours in the face of a subfreezing arctic wind. Over the millennial span of its evolution, the poor-will has adapted to these abrupt changes of season with the capacity of instant hibernation — an ultimate solution to the ultimate harshness of the desert. Ω

NATURE'S MARVELS OF ADAPTATION

The Namib Desert of southwestern Africa is one of the world's oldest and driest deserts. Consequently, the animal species that have learned to survive there have achieved the most thoroughgoing adaptations to desert life to be found anywhere. Amid evidence of a vanished Stone Age culture that succumbed to encroaching aridity, these animals serve as models of nature's ingenuity in the face of a scarcity of water and an abundance of heat.

Such remarkable adaptations as these require eons to accomplish. It takes many generations to test the value to the species of an accidental discovery or a genetic mutation, and then countless more generations for the success to spread. Humans, relatively recent arrivals on the planet, have been for the most part evaders of the desert, with neither the time nor the need to develop physiological adaptations to aridity. On the other hand, animals such as those shown here and on the following pages have lived in arid regions for millions of years; slowly and collectively, they have learned to do such things as distill their own water, as does the beetle at left, or have developed broad snowshoe-like feet that are fitted for travel over loose sand.

These animals — all residents of the Namib and the neighboring Kalahari Desert — are so specialized that one scientist saw in them "a sneak preview of the spectacular adaptations to aridity that I might see in more familiar deserts if I could return to Earth millions of years hence."

In order to collect an early-morning drink, a head-stander beetle bows to the moisture-rich fog rolling in from the Namib's Atlantic coast. Because the insect's body is still chilled by the night's low temperatures, water droplets condense on its back and roll down to its mouth.

Protectively shaded in desert hues, a young sand grouse sips water from its father's abdominal feathers. By drenching his absorbent plumage in a distant watering hole, the male sand grouse may carry as much as an ounce of water back to his young.

Alert for danger, a bat-eared fox guards her pups. The huge ears offer two advantages: They sharpen the fox's hearing and increase the animal's surface area, enhancing its ability to radiate excess body heat to the environment.

With a flash of its pink tongue, a gecko
cleans its lidless eyes. The almost-transparent
lizard stays under the sand during the
day but skims across it at night, aided by the
webbing on its broad feet. The shovel-like
feet also help the creature dig in to escape its
main predator, the stealthy snake at right.

A sidewinder viper scrawls its telltale
signature on Namibian sands. The snake's
unusual mode of locomotion counters the
tendency of loose sand to slip under its body;
poising on two points, it repeatedly rolls to
one side, marking the sand with set after set of
parallel imprints. Similar methods of
movement have been developed independently
by the sidewinder rattlesnake of North
America and by North Africa's sand adder.

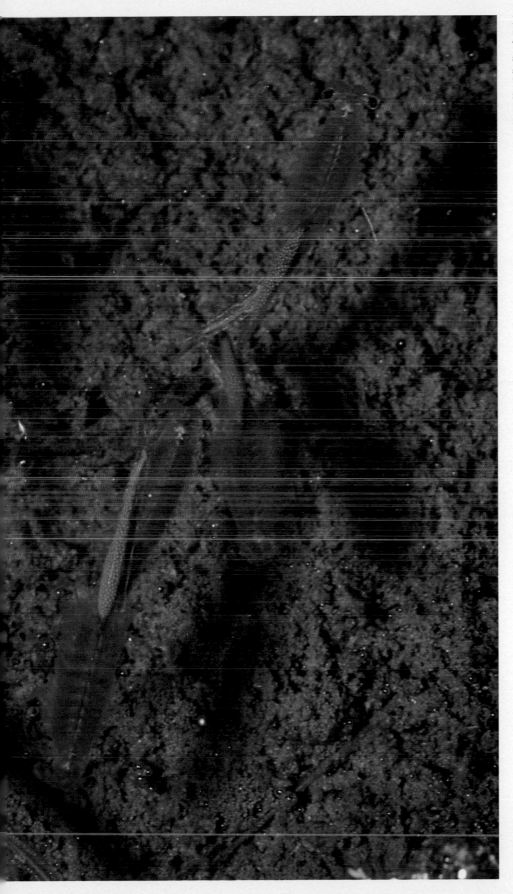

Flooded by a rare rainfall, a shallow pool in the Namib Desert teems with tiny fairy shrimp. These youngsters, only about 12 days old, already bear evidence of a new generation; eggs are visible through their transparent skin. When the water in the pool evaporates, the fertilized eggs may lie in the dried silt for 20 — perhaps even 100 — years before another rainstorm brings them to life.

In the absence of a plant or rock large enough to cast a shadow, a ground squirrel in the Kalahari Desert uses its plumed tail to create a shady spot for an afternoon snack.

A sidewinder viper's watchful eyes and forked tongue provide barely visible clues to the terror hidden beneath the sand. The snake's scales resemble sand in color and in texture, and its protruding eyes provide it with a clear view even when the reptile is almost completely buried.

STEMMING THE TIDES OF SAND

The earth's arid lands are spreading. Fluctuations in climate, both gradual and precipitous, along with other ecological shifts conspire to steal the moisture from large regions of the world. The process, called desertification, often is accelerated by a human factor: man's misuse of the fragile ecosystem at the edges of existing deserts.

Desertification's most visible symptom is soil erosion, a progressive degradation that continues in both dry and rainy seasons. During long dry periods, the soil cracks and disintegrates into particles that are blown about by local whirlwinds and by larger storms. Much of the fertile topsoil is carried away to clog water holes and accumulate into dunes of sand.

When rain finally comes to marginal lands, its blessings are mixed. Rain encourages the growth of vegetation, but it also causes a population explosion among herbivores — livestock, rabbits, rats. Grazing animals — and hordes of insects — crowd about the newly filled water holes and strip away all the nearby vegetation. Moreover, the force of falling rain can be as erosive as the wind. Wherever there is not enough vegetation with strong root systems to absorb the moisture, the rain carries away the soil, leaving behind a jagged landscape of empty rills and gullies.

When desertification threatens, humans tend to react in precisely the wrong way, overplanting and overgrazing in the desperate hope of producing enough to sustain them through the bad times. In the process they unwittingly contribute to the destructive cycle. Thus humans, existing precariously in a harsh environment, become both the primary villains and the ultimate victims of the spreading desert.

The problem is worldwide. Portions of nearly 70 countries — half of them in Africa alone — are afflicted by desertification. Each year 15 million acres of useful land are lost. The annual cost in productivity is $26 billion.

Because the problem is in large part caused by people, it is often within the power of people to avoid it. Some of the solutions are remarkably simple, but others are both complex and expensive. A United Nations study in 1980 estimated that reversing the tide of desertification by the end of this century could cost $90 billion.

For at least a century, scientists have known that the Sahara was expanding southward. The desert did not, as one might expect, advance along a broad front of dunes smothering everything in its path like a glacier of sand. Rather it spread like a pox; bare patches broke out here and there; vegetation withered, allowing the thin topsoil to blow away. The patches became larger and more numerous and eventually merged. In this way the Sahara

A turbaned African walks through the ruins of an acacia forest in Senegal, one of the sub-Saharan nations threatened by advancing desertification. Years of drought began the forest's decline; then starving animals stripped the leaves from the weakened trees, and humans chopped off branches for their fires.

A wind-blown tide of sand engulfs the huts of a village in Mauritania in 1981. More often, the desert advances less spectacularly, in patches like those afflicting the forest at right.

extended itself about four miles every year—a rate considered serious but not disastrous. Then the process accelerated to an apocalyptic pace.

In the late 1960s and early 1970s a devastating drought gripped much of the world. Nowhere was its impact more damaging than in the wide swath of semiarid land immediately south of the Sahara known as the Sahel. From the Atlantic eastward, the Sahel cuts across the boundaries of six recently founded nations that once were part of French colonial Africa. The region was one of scattered trees, scrub and seasonal grasses that supported a simple agricultural society.

The nomads who for centuries had made up most of the population of the Sahel raised cattle, camels, donkeys, goats and sheep; they roved over vast areas to find pasture for their herds. By contrast, the remainder of the population was distinctly immobile. They were farmers, tied to a few arable sites and dependent on the rains that fell between June and October to nourish their crops of cotton, beans, fruit trees, millet, sorghum and maize. Like most people in developing countries throughout the world, the inhabitants of the Sahel depended on wood for their fuel, gathering it from random patches of brush and from trees that could exist on little water.

134

It was, ironically, a succession of good years that betrayed the people of the Sahel. For half a dozen seasons in the 1960s, the monsoon rains fell more generously than usual. Encouraged, the hardy nomads challenged the desert, pushing farther north than before to graze their cattle. They built up their herds, proud symbols of wealth as well as sources of food. Behind them, the farmers expanded, tilling marginal land that traditionally had been suited only for grazing. International aid organizations helped the people to dig deep wells, enabling some cattle that had been kept continually on the move to linger in one place. At the same time, the introduction of modern health care to the region dramatically reduced the death rate, and the human population increased along with that of livestock.

After generations of bare subsistence, the Sahel seemed on the threshold of prosperity. In fact, however, it had never been so vulnerable. Beginning in 1968, the rains diminished and in 1970 they failed altogether. Drought was nothing new to the region; this was the third major siege in this century. Single dry years were common, but this time the years of insufficient rainfall came in a cluster, their destructive impact multiplying. The result was a human and ecological catastrophe.

Trade winds from the north and south meet near the Equator in an area called the intertropical convergence zone, causing clouds and updrafts. In normal years the zone shifts northward over the Sahel, bringing a rainy season that lasts through the summer. But in 1968, some small dislocation took place. Perhaps it was a weakening of the air currents in the intertropical convergence zone, or a cooling of the Atlantic waters that lowered the moisture content of the winds. In any event, for five years the rains in the Sahel were far below normal.

Nomads felt the pinch first. The land around their new wells was trampled and overgrazed until it was ripe for erosion. Their livestock soon devoured the remaining grass, then turned elsewhere to ease their appetites: Cattle stripped the trees while goats grubbed roots from the earth. As the dry years ground on, animals died by the thousands and eventually by the millions. In one nation, Upper Volta, not one cow in six remained alive.

Farmers, sorely overextended, saw their grain production fall by half. At last they were forced to eat their seeds, thus eliminating the possibility of planting a crop the next year, even as the land suitable for farming disappeared. Hunger worsened into famine. The nomads watched their last animals eat the woody branches of trees and shrubs in the absence of forage, killing trees that provided fuel for campfires and protection for the soil. Large numbers of the wandering, fiercely independent herdsmen surrendered their old way of life. They drifted south into the cities: Nouakchott, the capital of Mauritania, another of the six young nations in the Sahel, tripled its population. Huddled in refugee camps at the city's edge, almost 100,000 former nomads suffered years of epidemic, hunger and shame.

In some places the desert was now moving southward at 30 miles per year. Photojournalist Farrell Grehan, visiting the Sahel in 1973, described a storm in which "windblown dust covered everything. It stung the eyes, filtered through clothes, covered the body's pores and daubed every black, brown or white man the same dun-colored yellow. Here were grasslands without a blade of grass: once-fertile soil flung into the air, extinguishing the African sun."

At a remote clinic, Grehan watched as some nomads took a bundle off a camel's back. The bundle contained a tiny woman and her child. The two were given shots and put on the bare floor, the only space available. "During the second night," Grehan wrote, "the tiny woman died. The next morning I saw her child sitting alone in the clinic so covered with flies around his mouth and chin that he appeared to have a beard."

Although the outside world responded by sending motorized and even airborne caravans of food and medical supplies to the Sahel, a beard of flies on the chin of a suffering child became the indelible face of desertification. At least 100,000 people died; many others were permanently retarded by disease and brain damage traceable to malnutrition. The drought spread eastward, afflicting the Sudan and Ethiopia, two nations not previously considered part of the Sahel. Famine in Ethiopia contributed to a revolution in 1974 that evicted the long-established monarchy of Haile Selassie.

In 1973 and again in 1975 the rains returned to the Sahel, but they were sporadic and did not bring sufficient relief. What the nomads called the "drought with the long tail" continued. By the mid-1980s, the rains were still erratic. Few nomads ventured north to risk the desert. Desperate farmers attempted it, but to little avail. Pastoral nomadism was virtually at an end, the roving herdsmen now permanently on the dole, or existing on the fringe of their impoverished nations' urban society.

An accounting of the frightful cost of the drought in the Sahel must reckon not only with widespread suffering and death, but also with the consequent loss of invaluable knowledge. The retreat of the nomads from their parched domain in the Sahel, or anyplace else on earth where desertification prevails, means that the techniques of desert living no longer will be handed down from parent to child, as had been the case for many centuries.

Some of the earliest organized societies were roving bands of nomads who learned to cope with their hostile, arid environment while following the seasonal appearance of nuts and berries, or while searching out pasture for the first domesticated herds. Unlike their urban cousins, these mobile tribes kept no written records, yet evidence exists that their methods of survival have been passed on for millennia and are useful still.

At Tula'i on the Susiana Plain in Iran, an expedition led by American archeologist Frank Hole in 1973 discovered a rare nomadic archeological site. It is a harsh place where temperatures in the spring reach 120° F. and to touch the earth with one's foot is to raise a cloud of dust. Several members of the research team were former nomads, and they immediately saw much that was familiar in the 5,000-year-old campsite. The nomads recognized rectangles of stone as platforms used to keep bedding off the hot ground; they had used similar platforms themselves. Pacing a traditional two steps to the south, they dug and found the ashes of an ancient hearth. Nearby they found stones much like the ones they themselves used to batten down the edges of tents — tents very probably topped with woven goat hair like those of today, with openings prudently facing south, away from the prevailing winds.

At a site known as Choga Mami in equally arid Iraq, British archeologist Joan Oates in the early 1970s found one of the earliest systems of artificial irrigation. As early as 5500 B.C. people were growing barley and other crops at Choga Mami, which otherwise would have been too dry for farming. Oates found mounds of dirt remaining where ditches probably had

Cattle bones whiten in the sun amid the sparse scrub of the Sahel, where three million head of livestock were lost in the years of drought following 1968. Most of the cattle died not of thirst but of starvation after desertification stripped their grazing lands of forage.

been excavated to channel water to the fields from canals whose regular U-shaped sides and bottom were evidence that they were man-made.

Perhaps the most elaborate system of hydraulic engineering in the pre-Christian era was created in the rocky wastelands of what is now southern Israel. It was the work of a disparate collection of ancient nomadic bands who managed to settle down and prosper in the Negev desert. Their intricate network of stone terrace walls, conduits, catch basins and cisterns was basically so sound that, after centuries of disuse, parts of it have lately been rehabilitated and returned to operation *(pages 138-139)*.

For thousands of years, a much more rudimentary knowledge has sustained the aborigines of Australia: They simply know the whereabouts, in season, of everything they need in order to exist. Three quarters of their sprawling island continent is defined as arid; in the driest places rainfall averages only four inches a year. When Europeans began settling Australia in earnest about 1840, it was inhabited by an estimated 300,000 people who were divided into many small, independent tribes. Most of these aborigines clustered along the rivers and seacoasts, but a sizable number of them were dispersed through the parched interior area known as the Outback. By necessity the most mobile of peoples, they spent a large part of their time "chasing water." They had few if any clothes, no

From atop a limestone plateau *(background)*, the ruins of the ancient Nabataean town of Avdat look down on a partially reconstructed farm in Israel's Negev desert. Rain water coursing down wadis in the area's steep hills is deflected by a network of low walls *(right foreground)*. The water is diverted to terraced fields *(dark rectangles)* that are back in cultivation for the first time in centuries. On the small hill at left, a team of modern Israeli farmers has built a farmhouse and laboratory.

Restoring Ancient Desert Farms

The forbidding Negev desert that covers 60 per cent of modern Israel is the same "evil place" cursed by the nomadic followers of Moses as they wandered in search of the Promised Land. Yet there is strong archeological evidence that only a few centuries before Moses' time, in the hilly central highlands of the desert, six towns and thousands of farms flourished and a full half a million acres of land were in productive use.

The secret, subsequently lost to the ancients, was a system of hydraulic engineering now known as runoff agriculture. It consists of harvesting rain water from a large area of high ground and concentrating it on small terraced fields below. The Negev was a dense patch-work of such farms, with stone conduits delivering the precious water.

Israel's need to make the most of its limited water supply has not slackened over the centuries. In the 1950s a team of Israeli scientists, hoping to put the experience of their forefathers to modern use, set out to reconstruct a few of the old desert farms. The team determined that the earliest attempts to harvest the rain were low walls of stone built at right angles across the dry stream beds, called wadis, that sliced down the hills of the Negev. In the rainy season, when the wadis turn to racing floods, the walls slowed the cascades and also controlled erosion by building up arable soil behind each wall. Subsequently, conduits of loose rock were built to channel runoff water from the wadis and from huge catchment zones downhill to the farms. There the water was distributed to the walled fields or stored for a drier day in underground cisterns.

The runoff system enabled farmers then and now to nourish their land with three to five times as much water as natural rainfall provided. And the 20th Century researcher-farmers learned — just as their ancestors must have — that a mild slope provides more runoff than does a steep one. They theorize that either the skimpy soil on steep land does not shed water well when it is wet, or the prevalent outcroppings of rock offer more breaks for the water to infiltrate.

A restored wall of piled stone stretches across a hillside in the Negev, part of an ancient system intended to catch runoff water from above and channel it to fields below.

Puddles form after a rain in the earthen-walled peach orchard of a reconstructed farm in which each tree has its own small catchment basin. In the Negev, the runoff from 25 acres of land is needed on the average to provide enough water for one cultivated acre.

houses and no more possessions than they could carry with them on foot.

No movement of an aboriginal band was ever aimless or whimsical. In the flat, clear-skied Outback, the merest sprinkle of rain is visible to the experienced eye from 50 miles away, and the aborigines could relate the location of the rain to their knowledge of exactly where water catchments and edible plants existed. It was this keen awareness that guided each tribe's travels over several thousand square miles of seemingly trackless scrubland. This life of foraging resisted all but the worst of droughts. If no berries or nuts or edible roots were available at one site, the band simply moved on to another. Rarely were any provisions stored, because they could not be carried. But stored in the memory of even the youngest aborigine was a detailed map of the region in which the band moved.

Youngsters in the Outback learned the details of the territory they lived in by memorizing "dream-time" stories, religious tales about eternal be-

Armed with a digging stick, a young woman (*left*) of the Bushman society probes the sandy soil of southern Africa's Kalahari Desert for tubers. Another member of the nomadic tribe (*below*) squeezes a drink from one of the fleshy roots, a main source of water during the 300 or more days each year when not even a puddle is to be found in the Kalahari.

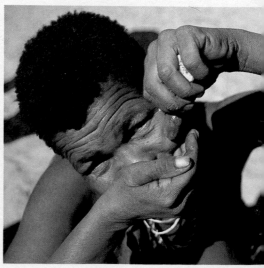

ings — part men and part lizards or kangaroos — who traveled the land, creating its surface features. A white streak in the rock might be described as the place where the kangaroo, in its first roamings on earth, dragged its tail. To an outsider it would have been a hardly noticeable feature in a monotonous landscape, but to an aborigine it was a reliable signpost.

The aborigines' most important secret was the ability to find water where none appeared to exist. William Tietkens, an early European explorer of the Australian interior, wrote of learning one of the aborigines' techniques in the nick of time. Tietkens' water bags were empty, and the day grew unbearably hot. Suddenly the aborigine boy accompanying Tietkens bolted toward a lone desert oak. He dug down at its base, snapped a horizontal root and pulled it away from the trunk, exposing about eight feet of root. The boy broke off a piece of root at the far end and held it over his mouth, and water drained into his throat. Gratefully, Tietkens imitated his guide.

The aborigines extracted water from another unlikely source — frogs. One species of desert frog adapted to the dry season by encasing itself in a cocoon filled with water and burrowing about a foot beneath the surface to hibernate until the rains came. The frogs, however, could be tricked. If an aborigine stamped his foot on the ground the noise might resemble the thunder that preceded rain. The sound sometimes provoked the frogs into croaking, giving away their whereabouts. The aborigines then probed the ground with digging sticks, snatched up the frogs and squeezed them like lemons. The water trickled into a waiting bowl or mouth.

The uncompromising requirements of survival in arid environments are the same everywhere. Thus the Bushmen, a society of hunter-gatherers in the Kalahari Desert of southern Africa, long ago learned to practice many of the same skills as the aborigines of Australia. Their material culture is somewhat more complex: Whereas the aborigines erect only grassy windbreaks for shelter, the Bushmen's windbreaks may be expanded into huts, and a circular camp of huts will arise in a place where food plants are plentiful. Wild plants and animals provide the bulk of the Bushmen's diet, and such plants as melons, tubers and succulents provide most of their water, since only in the rainy season is free-standing water available.

Like the aborigines, the Bushmen time their moves from place to place with a precise knowledge of the landscape and the seasonal availability of water and plant life. They camp in one place as long as it will sustain them. Then they simply abandon the camp and move on. The Bushmen have another technique for getting through the driest seasons: The band breaks up and individual families go off by themselves, sometimes changing camps a dozen times within a few months. This seasonal dispersal is based on the sound premise that many desert sites that cannot sustain an entire tribe will have enough food and water to keep a single family alive.

In spite of this reservoir of knowledge, the years of the nomads seem numbered. In the Sahel, drought and famine have driven them off the land into government resettlement programs. In other places, the intruder is civilization. Only a few thousand Kalahari Bushmen still hunt and gather in the simple, ancient way. Ranching and mineral exploitation have claimed much of the land, and most of the Bushmen have been absorbed into society at its lowest level. In Australia, the last nomads disappeared from the Outback in the 1960s. The aborigines now live uneasily in settlements or cities,

although in recent years some have returned to the countryside, establishing outstations that depend on the nearest settlement for supplies.

In the United States, at least one ancient people has withstood both the privations of a semiarid climate and the encroachment of civilization. They are the 10,000 Hopi Indians who farm the empty, scrub-covered plain of northeastern Arizona. The Hopi dwell in a dozen villages atop steep-walled sandstone mesas that are underlain with shale. Each year the region gets eight to 12 inches of rain. The water percolates through the sandstone and along tilted aquifers until it erupts in a handful of springs near the base of the Hopi mesas.

The Hopi are a deeply religious people who believe that their elaborate snake dances bring the precious rains, and every step in the raising of corn, their staple crop, is accompanied by ritual little changed for 10 centuries. But they supplement faith with a practical approach to farming that has enabled them to cultivate the dry washes below the mesas and produce fruits and vegetables in surprising abundance.

The natural springs, augmented by a few wells sunk into the aquifer, supply enough water for drinking, but never enough for irrigation. Precipitation comes in the form of light winter snow and occasional showers in spring and summer. Often the rain is extremely localized. Two Hopi farmers can stand in their cornfields, separated by only a few hundred yards, and watch a cloud scud across the plain and drop a brief shower precisely between them, missing both their fields. The rain that does find its mark is quickly absorbed into the sandy soil, leaving little or no standing water.

Illumined by the evening sun, the stone houses of a Hopi village cling to the top of a finger of Black Mesa in desolate northeastern Arizona. Only eight to 10 inches of rain falls here each year — and most of that in the winter — yet by making the most of subsurface water, the Hopi have successfully cultivated the lowlands around them for 1,000 years.

142

The Hopi have made the most of the available water by developing a system of dry farming (*page 165*). They establish fields, usually about 100 yards square, by clearing the brush and shrubs from dry washes or depressions in the sandy plain, where moisture is most likely to accumulate. From April to July they plant different fields in different varieties of corn — as well as melons, beans and squash. To plant corn, the farmer uses a digging stick to make a relatively deep hole, about 12 inches down, into which he drops a dozen kernels. The holes are placed a good four paces apart to give each plant sufficient earth from which to draw moisture. Simple windbreaks are built, to protect the young plants from being smothered by blowing sand. Weak shoots are weeded out early, and the plants send taproots deep into the ground beneath the sand in search of sustenance. The plants eventually resemble low shrubs and produce ears in three to four months.

Rain during the crucial growing period is the Hopi farmer's ally — but an untrustworthy one. A heavy rain too early can result in the rotting of the young plants. A downpour in the first month after germination can send a flash flood down the wash, sweeping the plants away. And a light shower in July may come too late to help corn planted in April. Little wonder that the Hopi seek help from their gods to ensure the success of their crops.

People, however primitive, who had proved they could cope with arid lands received new respect from social scientists in the wake of the drought that parched so much of the world in the early 1970s. That disaster more than any event before it focused attention on the inherent vulnerability of the planet. As a result, governments pledged support for campaigns to turn back the desert, and earth scientists sought a better understanding of the causes of desertification.

The key to desertification is the relationship between moisture and soil. Scientists have long been aware of a thin and delicate layer of tiny plants — algae, fungi, mosses and lichen — found in desert soils. Collectively they are called cryptogamic crust. Only recently has it become clear that these plant communities benefit the soil in a variety of ways. The short root hairs of mosses and lichens work to hold the soil in place. Filaments of algae and fungi growing just under the surface stabilize the soil. Some of the algae enrich the soil by absorbing nitrogen from the atmosphere and from the air

A modern adaptation to life in the desert, solar panels make efficient use of the desert sun to heat a home near Santa Fe, New Mexico. When temperatures fall below freezing at night, as they often do, heat collected by the panels and stored in tons of rock underground is admitted to the house through air vents.

in the soil. This fixing of nitrogen assists in the growth of other plants.

Together, the cryptogamic plants form a microscopic landscape of hummocks and valleys. This uneven terrain restricts the movement of standing water, allowing it time to seep into the ground rather than run off. The tiny plants also trap particles of silt that would otherwise blow past, and the silt serves to retain moisture.

This crucial coating of plant life is distressingly fragile. Exposure to trampling hoofs in overgrazed areas can wipe it out, as happened in the Sahel. With the cover gone, the soil's ability to retain water declines.

Climatologists generally agree that denuding the land generates what one of them calls a push toward desert conditions. They disagree, however, on the steps between. Some experts have sought the answer to the long drought in the earth's albedo — the degree to which it reflects the sun's heat. They contend that as land loses its vegetation, it reflects more heat back into the atmosphere, cooling the surface and bringing cold air downward, thus impeding rainfall. Others have contended that tiny soil particles rising into the atmosphere as dust from parched land tend to overseed the clouds, so that water droplets do not coalesce into raindrops large enough to fall. Conversely, it is argued that one result of the physiological activity of plants is the formation of solid particles that rise into the atmosphere and encourage the formation of raindrops. With vegetation gone, the particles are not available to seed the clouds.

In time, moisture returns to the air and even the worst drought ends. Sometimes the vicious cycle of desertification ends with it. The remedy can result from some climatic shift, or it may result from changes in the human use of marginal lands. The great drought of the 1930s turned the North American Great Plains into a dust bowl, threatening much of the region with desertification. The United States government, through its Soil Conservation Service, spent millions of dollars to reverse old, and harmful, agricultural practices, chiefly by introducing contour plowing and by planting trees as windbreaks. When the drought returned in less severe form in the 1950s and 1960s, the soil of the Plains states proved far less vulnerable. Some experts, however, warn that a future drought lasting several years could create another dust bowl.

Even such qualified success stories are few, and until recently they occurred mostly in those fortunate nations that have the resources to help an afflicted population through hard times and give long-term solutions a chance to work. In less-favored lands, the battle against desertification is an emergency that requires the commitment of meager government resources and the mobilization of legions of civilians whose futures are at risk.

Ethiopia is one of the poorest nations on earth, with annual per capita income of scarcely $100. Nine people in 10 live off the land, as subsistence farmers or herdsmen. The "drought with the long tail" continued to haunt the Ethiopians into the 1980s. Fully half the land in their country suffers from severe erosion. The forests are disappearing at the rate of 500,000 acres per year. Such figures are particularly distressing in a land that could be lush. Where it has not eroded, the topsoil of Ethiopia is fertile; the climate is relatively moderate except in the searing heat of the Rift Valley. In normal times, at least, there is enough rainfall to make the nation green. East of Addis Ababa, the capital city, the gentle hills of Asba Teferi are cool and fertile. But a steady growth in population has compelled people to

Making War on a Winged Plague

Since time immemorial, the ravenous desert locust has been the scourge of farmers and herdsmen across a swath of land 11 million miles square from the Atlantic shore of Africa to India's Thar Desert. Hordes of the flying insects, swarming on the wind, devour everything in their path, their acute sense of smell directing them unerringly to every leaf or blade of grass.

Once on the wing, locusts are unstoppable. But since the early 1980s, an international alliance of scientists has developed a way to combat the migratory pests where they are most vulnerable: in their breeding places. In order to breed successfully, the insect needs rain; its eggs will not survive without moisture, nor its young without budding greenery for food and shelter.

Images of the earth taken regularly from American and European satellites pinpoint the areas where moisture levels are favorable for locust breeding. The pictures are relayed to the Remote Sensing Office of the United Nations Food and Agricultural Organization, headquartered in Rome. Specialists there flash a warning to the nations about to be afflicted. As a result, local agricultural officials can concentrate their preventive efforts effectively, spraying the danger areas with pesticides to kill the locusts before they can proliferate.

A blizzard of desert locusts descends on children at play in the Ethiopian town of Keren during a particularly destructive outbreak in 1968. Related to grasshoppers, mature locusts have wingspans of up to five inches. The insects are voracious feeders *(inset)*, consuming their weight in food every day.

graze too many cattle on too few pastures, and to plant crops on slopes far too steep. Now the hills are marred by red gashes, dry gullies where the seasonal rains carry away more topsoil each year.

Despite a desperate shortage of funds and trained personnel, the Ethiopians began to fight back early in the 1980s. The government organized 20,000 peasant associations, consisting of 500 families each, to spend a portion of each week digging terraces and dams at Asba Teferi and other sites and learning the benefits of contour plowing and similar soil and water conservation techniques. The work requires many hands. It takes 230 person-days to dig one mile of terraces.

Planting trees is another means of slowing the desert's advance. To the limit of its resources, the Ethiopian government has planted or encouraged the planting of thousands of hardy eucalyptus and acacia trees, which protect the soil around them and fix nitrogen as well. The acacia also stays green year round and, if not overgrazed, can supply forage for cattle.

The problem of desertification is not Africa's alone. Each year immense clouds of hot dust rise over the Sahara and drift westward across the Atlantic. In 1982 a cloud more than 1,000 miles long reached Florida, dumping massive quantities of dust into the atmosphere along the way and raising air-pollution levels precipitously before it finally dissipated.

The United States in fact did not need a plume of Saharan sand to remind it of its own problems with advancing aridity. In the American West alone, 500 million tons of topsoil wash away into streams and rivers each year. By the early 1970s, the Bureau of Land Management reported: "There is very little of the western range where, because of the destruction of plant cover by improper management, accelerated erosion has not destroyed a portion of the soil mantle, and thus reduced the total productivity of the site."

Much of this destruction began a century or more ago, in the early days of the opening of the West. The use of public-domain rangelands began in earnest in the 1850s, and within 20 years the vast herds of livestock were overwhelming the land's capacity to support them. Two terribly harsh winters in the 1880s staggered the livestock industry and drove out many marginal ranchers. But the survivors continued to overgraze until, by 1932, the Western rangelands had lost an estimated 50 per cent of their original productivity. Smothered by unpalatable sagebrush, scarred by erosion and flash floods, large portions of 11 Western states were on the verge of becoming, in one observer's words, "a corrugated wasteland."

A slow turnaround began as the result of a combination of government and individual action. In 1934 the U.S. Congress passed the Taylor Grazing Act, which empowered the government to regulate the use of federal lands for grazing. Subsequently, itinerant herds were banned, and grazing allotments — specified as "animal units" — were awarded to range users. (A unit was defined as one cow or horse, or five sheep or goats.) Over the years, this policy has led to significant reductions in the size of herds, despite the fact that the Bureau of Land Management has too little manpower to police the range effectively; ranchers seem to have understood what is good for them and they honor the quotas voluntarily.

A related problem threatening the rangelands was the encroachment of brush. Sagebrush, rabbitbrush, mesquite and other useless but tough-rooted plants can invade a site and crowd out nutritious grasses and shrubs in the endless underground competition for moisture. An overabundance of

A wall of dust roiled by high winds closes in on the town of Alice Springs in Australia's arid interior. Such storms, which occur when a period of drought

has reduced normally arable soil to extremely fine particles, transport more than one half billion tons of dust worldwide each year.

brush makes land more vulnerable to erosion and at the same time inhibits its use. In 1973, the U.S. Department of Agriculture reckoned that three quarters of the grazing land in the Great Plains was covered by medium and dense stands of brush. About half of the mountain states and the Pacific states were similarly covered.

One of the regions most affected was the Vale district of southeastern Oregon. A place of rolling plateaus sliced by canyons, Vale had a history of overgrazing extending back to the middle of the 19th Century, when the Oregon Trail passed through. Partly as a result of this overuse, by the middle of this century almost 90 per cent of the district's vegetation had gone to sagebrush. Of the original cover of perennial grasses only a few patches remained. Much of the land was simply empty and barren. Worse yet, a weed called cheatgrass had invaded the dry, bare spaces around the sagebrush. The cheatgrass helped to prevent erosion but, as its name implies, it provided little forage—and indeed was harmful to animals. When gone to seed, the prickly plant catches in an animal's throat, causing great discomfort. The seed pods also collect in sheep's wool, reducing its value.

In the 1960s scientists from the University of California began a range-improvement program, experimenting with various methods of treating the land at selected sites in the Vale district. Some areas were sprayed with herbicides to reduce the sagebrush; others were sprayed and seeded in grass. Some were plowed and seeded, others treated with fire. Cooperating local livestock owners built fences, cattle guards, roads, wells and reservoirs to

In Iran, where oil is abundant and water scarce, workers try to stabilize shifting dunes by drenching them with a petroleum residue. The oily spray dries to a moisture-retaining mulch that enables shrubs and trees to take root.

control the movement of their stock — though not to reduce their numbers.

Spraying and seeding brought fast results. With the sagebrush reduced, wheat grass could compete with and reduce the density of cheatgrass. Erosion came to a virtual halt. Progress on test sites also made it possible to reduce overgrazing on sites that were not part of the test, so that they too began to return to good health. The scientists in fact showed that by intensively manipulating about 10 per cent of the region, they promoted favorable changes in the vegetation of the entire area. After more than a decade of work, the land was on its way, without much further human effort, to regaining its former balance: one quarter sagebrush, three quarters perennial grasses. The grazing capacity of the land had increased by half.

The Vale experiment demonstrated that endangered rangeland can be brought back. But there are reservations. The program was expensive, costing about $12 million. And a key element in its success was the use of certain strong herbicides that have become so controversial because they contain dioxin, a suspected cause of cancer, that the government has suspended their use.

Neither the problem of spreading sagebrush nor the danger of herbicides is as basic to the future of the American West as the availability of water. In much of the territory west of the 100th meridian, a line that runs from North Dakota down through west Texas, water is in short supply and its distribution radically uneven. The solution here has been massive irrigation. Between 1940 and 1980, irrigated acreage more than doubled in the

In an experimental project on a farm in Tunisia, a crop of spineless cactus grows alongside the thorny variety. The spineless cactus, planted on steep hillsides to stop erosion, also provides excellent dry-season fodder for livestock.

Organic dye used as a catalyst in the mining of potash adds a brilliant hue to the 400 acres of evaporation ponds at the Texasgulf mine on the Colorado River near Moab, Utah. River water is pumped through the ore deposits 3,000 feet belowground, then channeled to the shallow ponds. After a year of exposure to the desert sun, only crystals of potash remain.

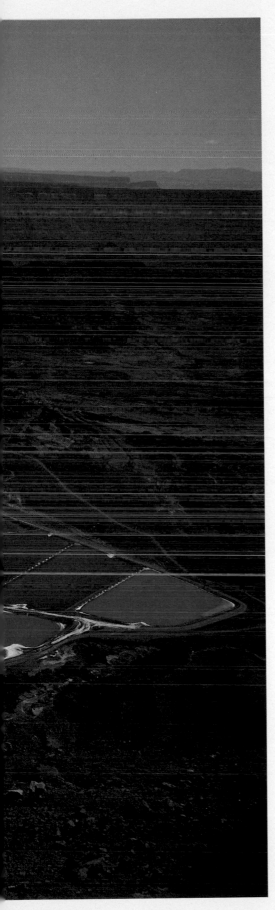

West; in one section of western Kansas the increase was a dramatic 3,000 per cent. As a result, today two thirds of U.S. crops are grown in the tier of Western states.

Innovations in technology give the modern irrigation project a far different look than the familiar system of dams and canals that has been used in one form or another for 7,000 years. Land graders guided by laser beams are employed to achieve the most efficient slope to the land. Huge rotating contraptions composed of pipes, wheels, pumps and valves are capable of spraying 130-acre circles at a time (pages 160-161). Some experimental sprinklers are controlled by computers to assure the proper timing, frequency and intensity of watering. They use water pumped out of the ground, since it is less likely to be harmfully salty than water that has been held behind a dam or sluiced in ditches through mineral-rich soils. The new sprayer systems can also be used in places where gravity will not help, and they permit fertilizers to be added to the water and sprayed with it.

But modern technology has not solved all the problems inherent in irrigation. In areas where underground water is most heavily used, the water table is dropping rapidly. Some of the water now being sprayed on American wheat fields is thousands of years old and will take thousands more to replace. In parts of Arizona, California, Texas, Nebraska, Kansas and Washington, the aquifers are drying up, being used at rates in excess of their replenishment. In many of these areas, salinization of the soil is also a major problem. Nutrient-rich minerals leach out of the soil and concentrate, rendering the soil infertile. When the fields are abandoned as a result, the usual outcome is wholesale erosion.

If poorly planned, irrigation projects can cause almost immediate ecological disaster. It has in fact been estimated that as much land is lost to faulty irrigation schemes as is regained by successful ones. There is an ironic form of desertification popularly called the wet desert, in which poor drainage leads to waterlogging of the soil; the irrigated area becomes soggy and saline, unfit for growing anything.

Dependable drainage, the key to avoiding such a problem, is expensive. It involves reshaping the land, moving millions of cubic feet of earth, building seepage-proof dams and lining irrigation canals with plastic or cement. But the cost of not investing in these initial steps can be calamitous. In India, an estimated 15 million ambitiously irrigated acres have recently become wet deserts, and 25 million more acres are threatened.

In part because of such failures, scientists are finding better ways to cope with arid lands; they are learning not to force upon dry regions techniques appropriate only to moderate climates, and they are learning to exploit the benefits marginal sites have to offer. In many dry countries, for example, where the forests dwindle and the populations expand, there will never again be enough wood to supply the housing needs of the people. A modern alternative, concrete, has proved too expensive and energy-intensive to be fully satisfactory. The answer has been a return to one of the few abundant commodities of arid lands: mud. Houses built of mud now shelter half of humanity. In the American Southwest more than 500,000 people live in homes constructed of adobe — mud bricks dried in the hot sun. Such houses are much in demand because they are both economical and comfortable: cool in summer and cozy in winter. The shelter for much of rural China is made of sun-dried earthen bricks, and similar mud construction is common across arid stretches of India and Africa where wood is unavailable and concrete too dear. The drier the climate the better; where rainfall and humidity are minimal, a mud house can last for 50 years or more.

Food, even more than housing, is the pivotal need in arid lands. Agronomists around the world have come to realize that, rather than try to adapt the desert to suit water-loving crops, the wise course is to find uses for existing desert plants, and to adapt new plants to desert conditions. Recent efforts to accomplish this are widespread—and the results are heartening, though not always commercially viable.

The roster of arid-zone plants found to have value as food or industrial products grows longer each year. Russian thistle, for example, is an aggressive weed that was introduced accidentally into South Dakota in 1873 and spread with dismaying persistence throughout the Western range. The thistle, which requires little water, is being evaluated as a forage crop. The plant is very high in protein and other nutrients, and its very aggressiveness and tenacity could be turned to advantage. It is also possible that saltbushes, a family of shrubs that thrives in dry, alkaline soil, could be used for forage—and since the bushes absorb salt, they offer the added benefit of helping to reclaim the land. Similarly, Egyptian scientists have found that a hardy rush called juncus, the seeds of which are rich in oils and proteins, not only grows well in saline soil but serves to reduce the soil's salinity. Creosote bush, long regarded as a useless pest in the American desert, is an excellent preservative and has been used experimentally to control the growth of cancer cells. Buffalo gourd, a wild plant of the Southwest, shows promise as a food crop, supplying edible oil, protein and carbohydrates.

Two of the best-known desert plants are jojoba and guayule. Jojoba is a shrub that grows wild in Mexico and in the Southwestern United States. It produces an oil-rich seed that has been touted as the savior of the endangered sperm whale, for jojoba oil is very similar to the whale oil that is in great demand as a lubricant for high-speed industrial machines. The oily wax from jojoba has a number of other applications, including use as an ingredient in shampoo, chewing gum, polishing wax and candles that burn brilliantly with no smoke. Jojoba is being studied carefully by half a dozen nations for another of its attributes: It thrives on marginal soils, needing no more than three inches of rain per year. In the first month after planting, the roots grow one inch a day, and a mature plant can have roots up to 100 feet long. This makes jojoba a prime weapon for planting on the margins of the Sahara, to help in the fight against desert creep.

Guayule, a shrub native to arid parts of Texas and Mexico, produces rubber. During World War II it supplied part of the United States' needs when the nation's overseas sources were cut off. Interest in guayule diminished after the War with the development of synthetic rubber, but it has recently received a great deal of attention from the U.S. Department of Agriculture and the rubber industry. Scientists have found a way to make the plant more productive by spraying it with a chemical that causes its rubber content almost to double.

A modest wildflower that blooms in the Arizona desert offers a potential substitute for petroleum in the manufacture of plastics. *Lesquerella palmeri*, commonly known as popweed, is a plant about a foot high that springs up after rains. Researchers from Lehigh University found that popweed seed contains an oil very similar to castor oil. It may be used to make such tough rubbery articles as gaskets, toys, automobile dashboards and heels for shoes.

Agricultural experimenters around the world are seeking to breed new strains of food plants that are drought-resistant and tolerant of salt. Perhaps their most dramatic breakthrough is a new cereal food crop called triticale, a hybrid of wheat and rye that can be baked into a rich brown bread. The product of years of breeding and crossbreeding in various parts of the world, triticale contains the same amount of protein as wheat (about 13 per cent), but the protein in triticale is of higher quality. The plant is more resistant to disease than wheat, and most important, it outperforms wheat significantly on marginal land. It has been found to be highly drought-resistant and its growth is not hampered by saline soils. The largest grower of triticale as of 1983 was the Soviet Union, with the United States next. The new staple also is grown in Canada, China, Argentina and parts of Europe.

More than anything else, hot, arid lands offer an abundance of sun and salt. The challenge to a generation of engineers has been to turn these potentially destructive elements to human advantage. One who succeeded was the Israeli engineer Lucien Bronicki. The fruit of Bronicki's 25 years of research in solar energy is a huge pond covering 70,000 square feet of desert near the Dead Sea at a place called Ein Bokek. The pond produces 35 kilowatts of electricity in the summer and 15 in the winter.

Technically, Bronicki's creation and others like it are called salt-gradient solar ponds. In each of them a layer of very salty water rests on the bottom,

Planted in pots watered by irrigation tubes, saltbush plants at the University of Arizona are exposed to varying levels of salinity as high as that of full-strength sea water. The experiment is part of an effort to expand the world's food supply by developing new strains of halophytes, edible plants that tolerate or even thrive on the salty conditions common in arid lands.

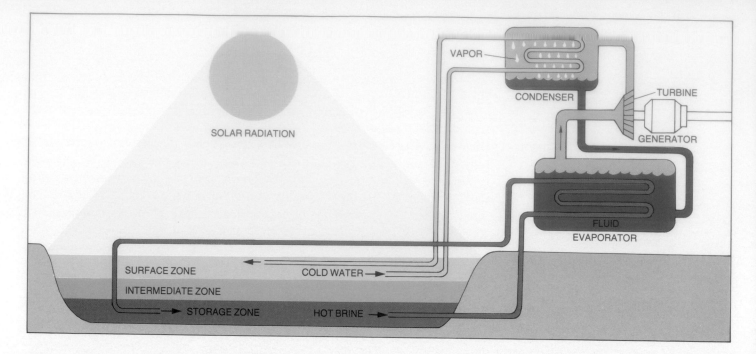

Diagram labels: SOLAR RADIATION, VAPOR, CONDENSER, TURBINE, GENERATOR, FLUID EVAPORATOR, SURFACE ZONE, INTERMEDIATE ZONE, STORAGE ZONE, COLD WATER, HOT BRINE

dense and heavy, while a layer of much fresher water lies on the surface. Between them is a zone of intermediate salinity that admits solar radiation but prevents heat from escaping. The sun penetrates the brackish top layers, and warms the brine below to the boiling point of pure water and even beyond. Normally, hot water, like hot air, expands and rises, but the lowest layer of a salt-gradient pond is so saturated with salt that it remains too heavy to rise to the top. Thus, the cooler, fresher and lighter water above acts as an insulating blanket, creating a natural heat reservoir, waiting to be tapped. Such ponds occur in nature, but they are rare because winds tend to roil the surface of salty ponds and mix the fresher water with the salt, reducing the difference between them to the point where the heat in the saltier water simply convects into the atmosphere.

In a man-made solar salt pond, such as the one at Ein Bokek, fresh water can be added to the surface level as it evaporates. Floating windbreaks and dikes keep the wind effect to a minimum, maintaining the reservoir of heat in the briny bottom water. Salt is added to the bottom layer to replace any that is diffused upward. It is relatively simple to draw the hot brine from the bottom of the pond and convert it into energy. In this way a pound of salt can in fact supply three times as much heat as a pound of coal; unlike coal, after the heat is extracted and used, most of the salt remains.

Potentially, the electricity made from solar ponds can be used to heat buildings, to run cooling systems or to power factories. Perhaps most promisingly, it can provide the power to desalinate water, through distillation or electrodialysis, making it suitable for drinking or irrigation in places where fresh water is in short supply.

The geography of the Middle East and northern Africa offers a number of natural sites for solar salt ponds. They are salinas, large salt flats or marshes that often are without standing water most of the year. Small canals could bring sea water to these places, which are often below sea level. Some of the water would be allowed to evaporate to create the especially salty lower level for the solar pond. Some of this brine would be put through a diffuser to create the less salty insulation layer. Unadulterated sea water would be used to form the least salty surface layer. A desalination process based on a salina 25 square miles in size could, by conservative estimate, produce 100 million cubic meters of drinking water per year.

A salt-gradient solar pond concentrates energy absorbed from the sun in the lowest of the pond's three zones of increasingly salty water. When hot brine from this storage zone is circulated through an evaporator, it vaporizes a fluid with a low boiling point. The vapor, flowing under pressure through a turbine, drives an electric generator, then is condensed back into liquid form by cool water pumped from the top of the solar pond.

The early-morning sun glints from the surface of a salt-gradient solar pond at Ein Bokek, on the shore of the Dead Sea. Plastic netting keeps the wind from churning the salt layers, and a long pipe carries hot brine from the bottom to a nearby evaporator, where the process of generating electricity begins.

Some ponds in dry areas of the United States are part of projects designed primarily to cope with large quantities of unwanted salty water. In Texas, the Army Corps of Engineers is constructing a 500-acre lake to hold brine springs and tributaries that would otherwise flow into the Red River, making it too salty for irrigation or drinking. A series of salt-gradient solar ponds within the lake will generate the electric power needed to pump brine into the lake from as far as 25 miles away.

In Israel, engineers building on the success of the solar pond at Ein Bokek have planned a massive hydroelectric project. It will operate on water from the Mediterranean, brought to the Dead Sea via a 140-mile-long canal and a 50-mile tunnel. The water, when allowed to pour into the Dead Sea (the lowest point on the earth's surface, at 1,312 feet below sea level), would run hydroelectric generators capable of producing 850 megawatts of electricity. The water would then be used to replenish the relatively unsaline upper layer of water in a series of large solar ponds to produce perhaps another 850 megawatts of electricity.

Making the sun and salt of arid lands work for humanity need not be a massive undertaking. From the Sahel to India to the Australian Outback, where the quest for firewood is a daily challenge, salt-gradient solar ponds offer an interesting alternative. Temperatures in solar ponds equal those in a slow-cooking pot, and dunking a watertight container of food into the pond would suffice to prepare a meal. The pond does not have to be large; one no bigger than a football field would amply serve a village of 100 people. A proliferation of such ponds, combined with techniques known and available to even the humblest of peoples, forms the arsenal with which humanity can change the course of its eternal struggle with the desert. Ω

THE GREENING OF THE WASTELANDS

Science has not yet learned to control the global forces that are inexorably expanding the deserts. But much progress has been made in the age-old battle to transform existing wasteland into productive soil. Indeed, 16 per cent of the land under cultivation in the world is arable only because it is irrigated. In the western United States, where a century ago a handful of Utah farmers nourished their meager crops with water carried across the desert on muleback, more than 40 million acres are now irrigated, and the management of water worldwide has come to involve high technology.

Many ancient systems are still in use, of course, particularly in the Middle East. Beneath the parched desert of Iran lies a 170,000-mile network of aqueducts, begun by the Persians as early as the Seventh Century B.C., that currently transports 75 per cent of the water used in Iran. But the ancient methods, by and large, are inefficient; according to some estimates, for each acre of desert transformed into productive soil, another acre was lost because of badly planned and executed projects. Evaporation has always been a major problem; moreover, inadequate drainage and leaking conduits could waterlog the soil and cause a toxic build-up of minerals.

Among the leaders in newer and better irrigation systems are the North African nation of Libya and the young state of Israel. Upon achieving independence in 1948, Israel moved swiftly to nationalize water as a vital natural resource. In the years since, the country has served as a laboratory for such techniques as drip irrigation, runoff collection wells and moisture-retaining plastic mulches. As a result, the 90 billion gallons of fresh water that flow through Israel's computer-regulated pipes and canals every year contribute to an agricultural bounty sufficient not only to meet the country's needs but to provide commodities for export.

Acres of green crops carpet a flood plain in an arid Moroccan valley. A system of canals wrests maximum advantage from the floods that enrich this low area with silt and nutrients.

158

Nurtured by sprinklers that rotate around a central pivot, saucers of wheat more than a half mile wide vitalize the Libyan Desert. Water drawn from as far

as 1,000 feet underground surges through nozzles on a length of pipe that travels on wheeled carriages.

Pipes draped with plastic sheeting dribble water along rows of crops in the Jordan Rift Valley. Drip irrigation, which was developed in Israel,

concentrates a slow, shaded trickle of water near the roots of the plants, while the plastic minimizes evaporation.

Greenery traces the path of an Iranian *qanat*, a channel that taps the water table at higher elevations to nurture fields in the lowlands.

By making the most of groundwater just beneath a dry Arizona riverbed, Hopi Indians raise corn by the so-called dry farming method.

Fertile fields cling to a desolate Himalayan slope in Pakistan, where because of the rain shadow created by the mountains there is almost no ground water.

An elaborate sluice-gated ditch system garners runoff from mountaintop rain and melting snow and routes it to the terraced fields.

A desert transformed, Arizona's Yuma valley — part of the Sonoran Desert — boasts more than 46,000 acres of productive farmland. The bounty is the

result of a massive diversion of water — more than 76 billion gallons a year — from the Colorado River.

ACKNOWLEDGMENTS

For their help in the preparation of this book the editors wish to thank: **In Australia:** Canberra — The Australian National University, Research School of Earth Science; Barbara Perry, The National Library of Canberra; Elwood — Ron Ryan, Photographic Agency of Australia; Kambah — Marguret Price; North Sydney — Cyndi Tebbel, The Photographic Library of Australia. **In France:** Auxerre — Henri-Jean Hugot; Nice — Jean-Paul Barry, Faculté des Sciences; Paris — Françoise Mestre, Jacana; Théodore Monod, Muséum National d'Histoire Naturelle. **In Great Britain:** Bristol — Joyce Williams; Kent — Brigadier R. A. Bagnold; London — Gillian B. Gibbins. **In Israel:** Jerusalem — Yehuda L. Bronicki, Ormat Turbines; Lieselotte Evenari; Professor Michael Evenari; Professor Aaron Yair, Hebrew University. **In Italy:** Florence — Paolo Graziosi, Fabrizio Mori, Università di Firenze; Rome — Jelle Hielkema, Jeremy Roffey, Plant Production and Protection Division, Food and Agriculture Organization of the United Nations; Instituto Italo-Africano. **In Sweden:** Stockholm — Bo Sommarström, Curator, The Sven Hedin Foundation, Ethnographical Museum of Sweden. **In the United States:** Arizona — (Flagstaff) Dr. Carol Breed, United States Geological Survey; (Tucson) Patricia Paylore, Office of Arid Lands Studies, The University of Arizona; Dr. Terah Smiley, Professor of Geosciences, University of Arizona; Thomas A. Wiewandt; California — (Hillsborough) Robert I. Gilbreath; (San Francisco) Edward S. Ross, California Academy of Sciences; Colorado — (Denver) Judith Totman Parrish, United States Geological Survey; New York — (New York) Nina Root, American Museum of Natural History; Texas — (Austin) Dr. Christopher Scotese, University of Texas. **In West Germany:** Braunfels — Helfried Weyer; Bonn — Professor Dr. Hanno Beck, Geographisches Institut, Universität Bonn; Hamburg — Susanne Schapowalow; Munich — Christine Hoffmann, Bayerische Staatsgemaeldesammlungen.

The index was prepared by Gisela S. Knight.

BIBLIOGRAPHY

Books

Adams, Robert, Marina Adams, Alan Willens and Ann Willens, *Dry Lands: Man and Plants*. St. Martin's Press, 1979.

Aleksandrova, V. D., *The Arctic and Antarctic: Their Division into Geobotanical Areas*. Cambridge University Press, 1980.

Andrews, Roy Chapman:
The New Conquest of Central Asia. American Museum of Natural History, 1932.
On the Trail of Ancient Man. G. P. Putnam's Sons, 1926.

Ashworth, William, *Nor Any Drop to Drink*. Summit Books, 1982.

Bagnold, R. A., *The Physics of Blown Sand and Desert Dunes*. London: Methuen, 1954.

Bailey, Liberty Hyde and Ethel Zoe, *Hortus Third*. Macmillan, 1976.

Baron, Stanley, *The Desert Locust*. Charles Scribner's Sons, 1972.

Barth, Dr. Heinrich, *Travels and Discoveries in North and Central Africa*, Vol. 1. Barnes & Noble, 1965.

Bertin, Leon, *Larousse Encyclopedia of the Earth*. Prometheus Press, 1961.

Birkeland, Peter W., and Edwin E. Larson, *Putnam's Geology*. Oxford University Press, 1978.

Blainey, Geoffrey, *Triumph of the Nomads*. Overlook Press, 1976.

Bodin, Svante, *Weather and Climate*. Dorset: Blandford Press, 1978.

Boorstin, Daniel J., *The Discoverers*. Random House, 1983.

Bowen, Ezra, and the Editors of Time-Life Books, *The High Sierra*. Time-Life Books, 1972.

Burton, Dr. Maurice, *Deserts*. London: Frederick Muller, 1974.

Clark, William R., *Explorers of the World*. Natural History Press, 1964.

Cleland, Robert Glass, *This Reckless Breed of Men*. Alfred A. Knopf, 1952.

Cloudsley-Thompson, John, *The Desert*. G. P. Putnam's Sons, 1977.

Colbert, Edwin H., *Evolution of the Vertebrates*. John Wiley & Sons, 1980.

Costello, David F., *The Desert World*. Thomas Y. Crowell, 1972.

Delpar, Helen, ed., *The Discoverers*. McGraw-Hill, 1980.

Denham, Major Dixon, Captain Hugh Clapperton and Doctor Oudney, *Narrative of Travels and Discoveries in Northern and Central Africa, in the Years 1822, 1823, and 1824*. London: John Murray, 1826.

Dickson, H.R.P., *The Arab of the Desert*. London: George Allen & Unwin, 1959.

Doolittle, Jerome, and the Editors of Time-Life Books, *Canyons and Mesas*. Time-Life Books, 1974.

Doughty, Charles M., *Travels in Arabia Deserta*. Random House, 1936.

Dregne, Harold E., ed., *Arid Lands in Transition*. American Association for the Advancement of Science, 1970.

Dunbier, Roger, *The Sonoran Desert: Its Geography, Economy and People*. University of Arizona Press, 1968.

El-Baz, Farouk, and Ted A. Maxwell, eds., *Desert Landforms of Southwest Egypt: A Basis for Comparison with Mars*. National Aeronautics and Space Administration, 1982.

Evenari, Michael, Leslie Shanan and Naphtali Tadmor, *The Negev: The Challenge of a Desert*. Harvard University Press, 1982.

Fletcher, W. Wendell, and Charles E. Little, *The American Cropland Crisis*. American Land Forum, 1982.

Flint, Richard Foster, and Brian J. Skinner, *Physical Geology*. John Wiley & Sons, 1977.

Gallant, Roy A., *National Geographic Picture Atlas of Our Universe*. National Geographic Society, 1980.

Gedzelman, Stanley David, *The Science and Wonders of the Atmosphere*. John Wiley & Sons, 1983.

Geology Today. Communications Research Machines, 1973.

George, Uwe, *In the Deserts of This Earth*. Harcourt Brace Jovanovich, 1976.

Glantz, Michael H., ed., *Desertification*. Westview Press, 1977.

Goetzmann, William H., *Exploration and Empire*. Alfred A. Knopf, 1971.

Goudie, Andrew, *The Human Impact: Man's Role in Environmental Change*. M.I.T. Press, 1981.

Green, Timothy, *The Restless Spirit: Profiles in Adventure*. Walker, 1970.

Hedin, Sven, *Through Asia*, Vol. 1. London: Methuen, 1899.

Herrmann, Paul, *The Great Age of Discovery*. Harper & Brothers, 1958.

Hoyt, William G., and Walter B. Langbein, *Floods*. Princeton University Press, 1955.

Hsieh, Chiao-min, *Atlas of China*. McGraw-Hill, 1973.

Hurlbut, Cornelius S., Jr., ed., *The Planet We Live On*. Harry N. Abrams, 1976.

Hutchinson, Sir Joseph, A. H. Bunting, A. R. Jolly and H. C. Pereira, *Resource Development in Semi-Arid Lands*. London: Royal Society, 1977.

Irving, Laurence, *Arctic Life of Birds and Mammals including Man*. Springer-Verlag, 1972.

Jackson, Donald Dale, and the Editors of Time-Life Books, *Sagebrush Country*. Time-Life Books, 1975.

Jaeger, Edmund C.:
Desert Wildlife. Stanford University Press, 1961.
The North American Deserts. Stanford University Press, 1957.

John, Brian S., ed., *The Winters of the World*. John Wiley & Sons, 1979.

Kirk, Ruth, *Desert — The American Southwest*. Houghton Mifflin, 1973.

Lamb, Edgar and Brian, *Colorful Cacti of the American Deserts*. Macmillan, 1974.

Lamb, H. H., *Climate, History and the Modern World*. Methuen, 1982.

Larson, Peggy, *Deserts of America*. Prentice-Hall, 1970.

Leopold, A. Starker, and the Editors of Time-Life Books, *The Desert*. Time-Life Books, 1961.

Ley, Willy, and the Editors of Time-Life Books, *The Poles*. Time-Life Books, 1970.

McGinnies, William G., and Bram J. Goldman, eds., *Arid Lands in Perspective*. American Association for the Advancement of Science and University of Arizona Press, 1969.

McGinnies, William G., Bram J. Goldman and Patricia Paylore, eds., *Deserts of the World: An Appraisal of Research into Their Physical and Biological Environments*. University of Arizona Press, 1968.

Maher, Ramona, *Shifting Sands: The Story of Sand Dunes*. John Day, 1968.

Mansfield, Peter, *The Arabs*. Penguin Books, 1982.

Milne, Lorus and Margery, *The Nature of Life: Earth, Plants, Animals, Man and Their Effect on Each Other*. Crown Publishers, 1970.

Moorehead, Alan, *Cooper's Creek*. Harper & Row, 1963.

National Geographic Society, *The Desert Realm: Lands of Majesty and Mystery*. National Geographic Society, 1982.

Norlindh, Tycho, *Flora of the Mongolian Steppe and Desert Areas*. Stockholm: Tryckrri Aktiebolaget Thule, 1949.

Page, Susanne and Jake, *Hopi*. Harry N. Abrams, 1982.

Perry, Richard, *Life in Desert and Plain*. Taplinger, 1977.

Pfeiffer, John E., *The Emergence of Society*. McGraw-Hill, 1977.

Philby, H. St. J. B., *The Empty Quarter*. London: Constable & Company, 1933.

Polo, Marco, *The Travels of Marco Polo*. Orion Press, 1958.

Pond, Alonzo W.:
The Desert World. Thomas Nelson & Sons, 1962.
Deserts: Silent Lands of the World. W. W. Norton, 1965.
Survival in Sun and Sand. W. W. Norton, 1969.

Press, Frank, and Raymond Siever, *Earth*. W. H. Freeman, 1978.

Readers's Digest, *Scenic Wonders of America*. Reader's Digest Association, 1980.

Rugoff, Milton Allan, *The Great Travelers*. Simon & Schuster, 1960.

Sheffield, Charles, *Earth Watch*. Macmillan, 1981.

Sheridan, David, *Desertification of the United States*. Council on Environmental Quality, 1981.

Short, Nicholas M., Paul D. Lowman Jr., Stanley C. Freden and William A. Finch Jr., *Mission to Earth: Landsat Views the World*. National Aeronautics and Space Administration, 1976.

Strahler, Arthur N., *The Earth Sciences*. Harper & Row, 1963.

Sturt, Captain Charles, *Narrative of an Expedition into Central Australia*, Vol. 2. Greenwood Press, 1969.

Sutton, Ann and Myron, *The Life of the Desert*. McGraw-Hill, 1966.

Tanaka, Jiro, *The San: Hunter-Gatherers of the Kalahari*. Tokyo: University of Tokyo Press, 1980.

Thesiger, Wilfred, *The Last Nomad: One Man's Forty Year Adventure in the World's Most Remote Deserts, Mountains and Marshes*. E. P. Dutton, 1980.

Thomas, Bertram:
Alarms and Excursions in Arabia. Bobbs-Merrill, 1931.
Arabia Felix: Across the "Empty Quarter" of Arabia. Charles Scribner's Sons, 1932.

Thurman, Harold V., *Introductory Oceanography*. Charles E. Merrill, 1981.

Trench, Richard, *Forbidden Sands: A Search in the Sahara*. Academy, 1982.

Viking Lander Imaging Team, *The Martian Landscape*. National Aeronautics and Space Administration, 1978.

Wagner, Frederic H., *Wildlife of the Deserts*. Harry N. Abrams, 1980.

Walker, Ernest P., et al., *Mammals of the World*, Vol. 2. Johns Hopkins University Press, 1975.

Warburton, Colonel Peter Egerton, *Journey Across the Western Interior of Australia*. London: Sampson Low, Marston, Low, & Searle, 1875. Reproduced by Libraries Board of South Australia, Adelaide, 1968.

Woodin, Ann, *Home Is the Desert*. Macmillan, 1964.

Wulff, Hans E., *The Traditional Crafts of Persia*. M.I.T. Press, 1966.

Younghusband, Captain Frank E., *The Heart of a Continent: A Narrative of Travels in Manchuria, across the Gobi Desert, through the Himalayas, the Pamirs, and Chitral, 1884-1894*. London: John Murray, 1896.

Periodicals

Clark, Wilson, "Take Sun and Salt, Add Some Water — the Result Is Energy." *Smithsonian*, October 1980.

Cornejo, Dennis, "Night of the Spadefoot Toad." *Science '82*, September 1982.

Edesess, Michael, "On Solar Ponds: Salty Fare for the World's Energy Appetite." *Technology Review*, November/December, 1982.

El-Baz, Farouk, "Desertification." *Smithsonian*, June 1977.

Elston, D. P., and S. L. Bressler, "Paleomagnetic Investigation of Cenozoic Glaciogenic Sediments, Taylor Valley and McMurdo Sound." *Antarctic Journal of the United States*, 1980 Review.

Englebert, Victor, "The Danakil: Nomads of Ethiopia's Wasteland." *National Geographic*, February 1970.

Gilbreath, Robert I., "Tracking Desert's Tiniest Flowers." *Smithsonian*, April 1974.

Gore, Rick, "The Desert: An Age-Old Challenge Grows." *National Geographic*, November 1979.

Hamilton, William J., III, "The Living Sands of the Namib." *National Geographic*, September 1983.

Long, Michael E., "Baja California's Rugged Outback." *National Geographic*, October 1972.

McGinnis, L. D., "Seismic Refraction Studies in Western McMurdo Sound." *Antarctic Journal of the United States*, 1980 Review.

Morris, C. E., L. Knopoff, P. A. Rydelek and W. D. Smythe, "Observations of Free Modes, Tides, and Tilts at Amundsen-Scott Station." *Antarctic Journal of the United States*, 1980 Review.

Nagata, T., "Earth Sciences Research in the McMurdo Sound Region, 1979-1980." *Antarctic Journal of the United States*, 1980 Review.

Page, Jake, "The Improbable Wild Gardens of Our Deserts." *Smithsonian*, April 1972.

Ross, Edward S.:
"The Ancient Namib Desert." *Pacific Discovery*, July/August 1972.
"Introducing the Desert." *Pacific Discovery*, March/April 1952.

Shabtaie, S., C. R. Bentley, D. D. Blankenship, J. S. Lovell and R. M. Gassett, "Dome C Geophysical Survey, 1979-80." *Antarctic Journal of the United States*, 1980 Review.

Shoji, Kobe, "Drip Irrigation." *Scientific American*, November 1977.

Splinter, William E., "Center-Pivot Irrigation." *Scientific American*, June 1976.

Stump, E., "Two Episodes of Deformation at Mt. Madison, Antarctica." *Antarctic Journal of the United States*, 1980 Review.

Tasch, P., "New Nonmarine Fossil Links in Gondwana Correlations and Their Significance." *Antarctic Journal of the United States*, 1980 Review.

Thornes, J. B., "Dynamic Earth 1: On the Threshold of Landscape." *Geographical Magazine*, November 1983.

Tijmens, Willem J., "From an Ancient Desert Relict." *Natural History*, April 1967.

Toufexis, Anastasia, "The Sahara's Buried Rivers." *Time*, December 6, 1982.

Treves, S. B., "Hyaloclastite of DVDP 3, Hut Point Peninsula, Antarctica." *Antarctic Journal of the United States*, 1980 Review.

"Unveiling the Sahara's Hidden Face." *Science News*, June 26, 1982.

Whitson, Martha A., "The Roadrunner: Clown of the Desert." *National Geographic*, May 1983.

Wright, T. O., "Sedimentology of the Robertson Bay Group Northern Victoria Land, Antarctica." *Antarctic Journal of the United States*, 1980 Review.

Wulff, Hans E., "The Qanats of Iran." *Scientific American*, April 1968.

Young, Gordon, "The Essence of Life: Salt." *National Geographic*, September 1977.

Other Publications

Dictionary of Geological Terms. Anchor Press, 1976.

Encyclopedia of World Rivers. Rand McNally, 1980.

Goodin, J. R., and David K. Northington, eds., *Arid Land Plant Resources*. International Center for Arid and Semi-Arid Land Studies, Texas Tech University, July 1979.

Henning, D., and H. Flohn, *Climate Aridity Index Map*. Food and Agriculture Organization of the United Nations, the United Nations Educational, Scientific and Cultural Organization, and the World Meteorological Organization, 1977.

More Water for Arid Lands: Promising Technologies and Research Opportunities. National Academy of Sciences, 1974.

Parrish, Judith Totman, A. M. Ziegler and Christopher R. Scotese, *Rainfall Patterns and the Distribution of Coals and Evaporites in the Mesozoic and Cenozoic. Paleogeography, Paleoclimatology, Paleoecology*. Amsterdam: Elsevier Scientific Publishing Company, 1982.

Robinson, G. D., and Andrew M. Spieker, eds., *Nature to Be Commanded: Earth-Science Maps Applied to Land and Water Management*. U.S. Government Printing Office, 1978.

Sherratt, Andrew, ed., *The Cambridge Encyclopedia of Archaeology*. Scarborough, Ontario: Prentice-Hall of Canada/Cambridge University Press, 1980.

Smith, David G., ed., *The Cambridge Encyclopedia of Earth Sciences*. Crown Publishers/Cambridge University Press, 1981.

Spineless Cactus. Rome: Food and Agriculture Organization of the United Nations, no date.

PICTURE CREDITS

The credits for the illustrations that appear in this book are listed below. Credits from left to right are separated by semicolons, from top to bottom by dashes.

Cover: Gary Ladd. 6, 7: © Jim Brandenburg. 8, 9: Daniele Pellegrini, Milan. 10, 11: Schapowalow/Scholz, Hamburg. 12, 13: Kevin Schafer. 14, 15: Daniele Pellegrini, Milan. 16, 17: Stephen J. Krasemann/DRK Photo. 18: © Robert Frerck, Odyssey Productions. 20: Library of Congress. 22: National Archives. 25-27: Victor Englebert, Cali, Colombia. 28: Map by Bill Hezlep. 29: From *Through Asia*, by Sven Hedin, published by Methuen & Co., London, 1899, courtesy Library of Congress. 30: Map by Bill Hezlep. 31: Norbert Nordmann, courtesy Universitätsbibliothek, Bonn. 32: The Faculty of Oriental Studies, Cambridge. 34, 35: Royal Geographical Society, London. 36, 37: © Wilfred Thesiger 1976, London. 38: Map by Bill Hezlep. 39: National Library of Australia, Canberra. 40, 41: From *The New Conquest of Central Asia*, by Roy Chapman Andrews, published by The American Museum of Natural History, New York, 1932, except middle, courtesy of the Library Service Department, American Museum of Natural History. 42-47: Courtesy of the Library Services Department, American Museum of Natural History. 48: © Gordon Wiltsie 1980/AlpenImage. 51. JPL/NASA. 54, 55: © Dr. Georg Gerster/Photo Researchers. 58, 59: John A. Pawloski; Rod Borland/Bruce Coleman Inc. 60, 61: Peter L. Kresan. 63: Giorgio Lotti, Milan — drawings by Walter Hilmers Jr. 64: Marion and Tony Morrison, Suffolk. 65: Wardene Weisser/Ardea, London. 66: © Earth Satellite Corporation 1979. 68, 69: Peace River Films Inc. 70, 71: Kazuyoshi Nomachi, Tokyo. 73: Stephenie S. Ferguson, © William E. Ferguson — drawing by Walter Hilmers Jr. 74: Michael E. Long, © National Geographic Society — Dr. Georg Gerster, Zumikon/Zurich. 75: Dr. Georg Gerster, Zumikon/Zurich — The Photographic Library of Australia, Sydney. 76: © Dr. Georg Gerster/Photo Researchers. 77: © Earth Satellite Corporation 1980. 78: JPL/NASA. 79: Dr. Farouk El-Baz. 80-89: Maps by Bill Hezlep — drawings by William J. Hennessey Jr. 90: © 1983 T. A. Wiewandt and Scott Altenbach. 92: Robert I. Gilbreath. 94, 95: © T. A. Wiewandt 1981. 96: G. R. Roberts, Nelson, New Zealand. 97: Dan McCoy/Black Star; John Gerlach. 98: Kent and Donna Dannen — C. Allan Morgan. 100: © Gordon Wiltsie 1983/AlpenImage. 101-103: Edward S. Ross. 104, 105: © Galen Rowell 1983/High & Wild Photography. 107: Drawing by Susan Johnston. 109: C. Allan Morgan — © T. A. Wiewandt 1981. 110, 111: © T. A. Wiewandt 1981. 112, 113: Klaus Paysan, Stuttgart. 115: © Jean-Philippe Varin/Jacana, Paris — Ste-

INDEX

174